THE LITTLE BOOK OF
AVIATION

THE LITTLE BOOK OF
AVIATION

NORMAN FERGUSON

Front cover design: iStock Photo.
Back cover image: Red Arrows. (Chris Chidsey, stock.xchng)

First published 2013
Reprinted 2015

The History Press
The Mill, Brimscombe Port
Stroud, Gloucestershire, GL5 2QG
www.thehistorypress.co.uk

British Library Cataloguing in Publication Data.
A catalogue record for this book is available from the British Library.

ISBN 978 0 7524 8837 0

Typesetting and origination by The History Press
Printed and bound in Great Britain by TJ International Ltd.

CONTENTS

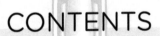

Acknowledgements 7

Introduction 8

The Little Book of Aviation 9

Bibliography 143

ACKNOWLEDGEMENTS

Any errors are the fault of the author and not in any way of those he consulted. Thanks go to the following individuals for their help and contributions: Raymond A. Ferguson; Derek N. Ferguson; Henrietta B. Ferguson; Des Brennan; Sandy Buchanan; Captain James A. Lovell Jr; Dick Rutan; John Farley; Jake Jarron.

The following online sources were used: RAF Museum; Royal Air Force; Little Friends; Aviation Safety Network; HistoryNet.com; Flight Global, *Flight* magazine archive; *Detroit Free Press*; *USA Today*; *Providence News*; *The Log*; *The Times*; US Centennial of Flight Commission; Civil Aviation Authority; SITA; IATA; United Airlines Historical Archive.

INTRODUCTION

In the 110 years since the Wright Brothers made their first powered and barely controlled flights at Kitty Hawk, the mode of transport they helped develop has transformed the way we view the world. Journeys that took weeks now take hours. Sights that were previously unavailable are now commonplace. Things that were not even thought of happen every day. To travel by air is now nothing special; a sign of how much aviation has progressed since those early pioneering days of risk and daring innovation.

Aviation is a wide-ranging subject and it is not difficult to find stories and facts that are inspiring, tragic, humorous, perplexing, impressive and many other things besides. Hopefully *The Little Book of Aviation* captures some of these.

Norman Ferguson
May 2013

✈ FIRST FLIGHT

120 – 175 – 200 – 852:
The distances in feet of four successive flights made on the most notable date in aviation: 17 December 1903. On this day, Orville and Wilbur Wright took the 'giant leap' that led to a revolution in transportation. The brothers from Ohio had worked on developing gliders before moving on to creating powered machines. On the morning of the 17th, on the beach at Kitty Hawk in North Carolina, Orville took his position on top of the lower wing of the Wright Flyer. With a photographer standing by, the biplane moved off down metal tracks before lifting off. Orville described the flight as 'exceedingly erratic' with the aircraft difficult to control and subject to large pitching movements. Nonetheless, it flew for twelve seconds, enough to complete the first powered and sustained flight by a heavier-than-air flying machine. Flights continued but on the fourth and last flight of the day, which lasted almost a minute, the machine hit the ground heavily and became damaged. While it was being retrieved, winds caused further damage. It never flew again. It was repaired for display and after a period in the Science Museum in London was returned to America. It is displayed in the Smithsonian National Air and Space Museum in Washington DC, alongside Charles Lindbergh's *Spirit of St Louis*, the sound barrier-breaking Bell X-1 and Apollo 11's command module *Columbia*. Apollo 11, in fact, carried small pieces of fabric and wood from the Wright Flyer to the moon.

Wilbur watches as his brother Orville Wright makes history at Kitty Hawk on 17 December 1903 by lifting off in the Wright Flyer. *J.T. Daniels, USLSS/US Coast Guard*

✈ FIRST SUPERSONIC AIRLINER

The first supersonic airliner was not Concorde, as might be thought. In 1961 another type was first to break the sound barrier. Flying from the location for so many aviation firsts – Edwards Air Force Base in California – a four-engined Douglas DC-8 was put into a dive from 52,000ft. The intentional test flight saw the airliner reach Mach 1.012 (660mph) before pulling up. No harm or damage was done and the actual aircraft later entered service with Canadian Pacific Air Lines.

✈ FAMOUS FEATS OF AVIATION: ATLANTIC THERE AND BACK

The Scottish county of East Lothian is known for its scenic golf courses, historic castles and one of the biggest gannet colonies in the world at the Bass Rock. What's less known is its place in aviation history. In the early hours of 2 July 1919 the biggest airship in Britain left its hangar at the airfield at East Fortune. The 643ft-long craft soon took off and headed west. After a journey of four and a half days that encountered poor weather and engine problems the dirigible landed in the USA. The R34 had completed the first east-to-west aerial crossing of the Atlantic. It touched down with approximately one hour's fuel left.

Along the way two stowaways had been discovered, a kitten called Wopsie and a human called William Ballantyne – a crew member who had been removed to make room for an American observer but didn't want to miss out. He was found over water, otherwise he would have been given a parachute and sent homewards. A parachute was used by one of the officers who jumped to help the American reception personnel who were unused to dealing with an airship of that size.

The crew were fêted by the people of New York, and met the American President Woodrow Wilson. After several days of being entertained and re-equipping the airship, it was time to return. The journey home encountered no major issues. The R34 was scrapped in 1921 following an accident. In the Museum of Flight that now stands on the East Fortune airfield site, the airship's nose cone, in the shape of a heraldic crest, can be seen.

✈ AVIATION-RELATED PUBS AND RESTAURANTS

Air Balloon Tavern, Bristol
Airfield Tavern, Yeovil
Airport Tavern, Bristol
Dambusters Inn, Scampton
Happy Landing, Stanwell
Memphis Belle, Westbrook
Red Arrow, Lutterworth
The Air Balloon, Birdlip
The Airfield, Hatfield

The Airman, Feltham
The Aviator, Cheltenham
The Barnes Wallis, North Howden
The Concorde, Rainham
The Douglas Bader, Martlesham Heath
The Flying Fortress, Bury St Edmunds
The Tiger Moth, Brickhill
The Vulcan Arms, Sizewell Beach
Whittle Inn, Hucclecote

✈ BATTLE OF 'BRITAIN'?

Australia, Barbados, Belgium, Canada, Czechoslovakia, France, Great Britain, Ireland, Jamaica, New Zealand, Poland, Rhodesia, South Africa, USA.

Countries who provided Allied aircrew that took part in the Battle of Britain.

✈ WARTIME SLANG

ack-ack – anti-aircraft fire
angels – altitude, e.g. 'angels one five' = 15,000ft
Archie – anti-aircraft fire
bandit – enemy aircraft
beat up – fly low-level over an airfield
prang – a crash
crate – aircraft
deck – the ground
drink – area of water
flap – situation involving panic
goon – German prisoner-of-war guard
kite – aircraft
Mae West – inflatable life vest
piece of cake – easy thing to do
scramble – rapid take-off
squirt – burst of machine-gun fire
Tail End Charlie – rear gunner
wizard kite – great aeroplane

✈ BOWLED OVER

William Gilbert Grace is regarded as one of the greatest cricketers ever to have played the game. Easily identified by his bushy beard and portly physical shape he was the most famous sportsman of his day. 'W.G.' ran up a long list of achievements, including 2,876 first-class wickets as a bowler and 54,896 runs with the bat. He was the first to score a hundred hundreds. Grace played cricket for forty-eight years and his last game was in July 1914 when he was 66 years old. His life without cricket was not to last too long, however. He became irritated with the Zeppelin raids that were flying above his Kent home and it was reported he would show his annoyance by shouting and shaking his fist at them. On the last Zeppelin raid of 1915, on the night of 13/14 October, fifty-five people were killed in London and the south-east of England. Grace suffered a heavy fall. He died of a heart attack ten days later.

# *TOP GUN* AIRCREW NICKNAMES

Characters in the film:
Charlie, Chipper, Cougar, Goose, Hollywood, Iceman, Jester, Maverick, Merlin, Slider, Sprawl, Stinger, Sundown, Viper, Wolfman.

US Navy F-14 aircrew, Top Gun instructors and 'MiG' pilots involved in making of the film:
Bio, Boa, Bozo, Circus, Curly, D-Bear, Flex, Hollywood, Horse, Jambo, Jaws, Loner, Organ, Player, Rabbi, Rat, Secks, Silver, Sobs, Squire, Sunshine, Tex, Too Cool, Vida.

# HOW TO FLY AEROPLANES

Rules of the Air:
'When flying in the air the pilot must obey certain rules like the motorist on the road who must keep to one side. If an aeroplane has to wait over an aerodrome, whether for room to land or for another aeroplane to join it in the air, it must circle to the left all the time. If it cannot wait, it must make a distress signal – perhaps fire a pistol – and it is then allowed to land.

'The pilot always moves on a left-hand turn unless he is more than 2 miles from the aerodrome, or more than 6,000 feet above the ground. The faster aircraft give way to the slower ones. When meeting or overtaking in the air, each aeroplane must alter course to the right, and no aircraft must dive to pass. No aeroplane must pass within 200 yards of any other and in the case of regular air routes this distance is 300 yards.'

(*The Book About Aircraft*, published 1933)

✈ DE HAVILLAND MOTHS

The de Havilland company built a variety of aircraft including the 'Wooden Wonder' Mosquito during the war and then later the world's first jet airliner, the Comet. In the pre-Second World War days, however, it was known for its range of light aircraft, particularly those named after a certain species of insect:

Cirrus Moth
Fox Moth
Genet Moth
Giant Moth
Gipsy Moth
Hawk Moth
Hermes Moth
Hornet Moth

Leopard Moth
Menasco Moth
Metal Moth
Moth Major
Moth Minor
Puss Moth
Swallow Moth
Tiger Moth

One of the famous Moths built by de Havilland was the Gipsy Moth. This example was built in 1929, and was one of 688 built. *Derek N. Ferguson*

✈ TOP TEN ACES OF THE SECOND WORLD WAR

Ranking	Name	Country	Number of kills
1	Erich Hartmann	Germany	352
2	Gerhard Barkhorn	Germany	301
3	Günther Rall	Germany	275
4	Otto Kittel	Germany	267
5	Walter Nowotny	Germany	258
6	Wilhelm Batz	Germany	237
7	Erich Rudorffer	Germany	224
8	Heinrich Bär	Germany	220
9	Hermann Graf	Germany	212
10	Heinrich Ehrler	Germany	208

The Luftwaffe pilots flew many more missions than their Allied counterparts. The highest-scoring Allied pilot was a Russian, Ivan Kozhedub, who shot down sixty-two enemy aircraft. Kozhedub flew 330 missions, compared to Hartmann's 1,400+.

WORLD'S FIRST FLIGHT ATTENDANT

In March 1912, Heinrich Kubis became the first flight attendant when he served passengers on the German airship *Schwaben*. He went on to serve as chief steward on the *Graf Zeppelin* and the *Hindenburg*, and was on board that ill-fated airship when it caught fire in 1937. Kubis managed to escape after first helping his passengers evade the inferno. He sat on a window ledge until the burning airship descended close enough to the ground to allow them all to jump. Kubis had also evaded injury when working on the *Schwaben* when it caught fire in 1912.

✈ GREAT PLANES: SPITFIRE

Of all the warplanes that Britain has produced, it is Supermarine's Second World War fighter that is the best loved and most easily identified. Its slim elliptical wings and neat fuselage, ending with a smooth, rounded tail – all pulled through the air by the Merlin engine with its classic engine sound – make the Spitfire stand out from its contemporaries. Its legendary position was not always assured. The design was difficult to produce and caused the

Reconnaissance Supermarine Spitfire PR.11s were unarmed and relied on speed and height for their defence and brought back much valuable information from enemy territory. *Derek N. Ferguson*

pre-war Air Ministry to consider cancellation. In the manner of many other high-tech projects, it was over budget. Fortunately the project was allowed to proceed.

The aircraft had a narrow undercarriage track that made it more problematic to land than the sturdier Hawker Hurricane. Come the two planes' 'finest hour' – the Battle of Britain – there were considerably more Hurricanes in squadron service than Spitfires (thirty squadrons had Hurricanes while only nineteen were equipped with Spitfires). However, it was the sleeker fighter the press and public came to know better. Luftwaffe pilots who were shot down by Hurricanes preferred their conqueror to be a Spitfire as they had a higher regard for it, tasked as it was to shoot down German fighters, while the slower Hurricane was directed to attack bombers.

The Spitfire was built right through the war and over 20,000 were made. It was used for fighter interception, photo reconnaissance and ground attack, while a sea-going version – the Seafire – was built for the Royal Navy.

 MACROBERT'S REPLY

Since 1982 each aircraft operated by the RAF's XV Squadron assigned the code letter 'F' has also been marked with the words 'MacRobert's Reply'. This tradition goes back to the Second World War, when two Short Stirling heavy bombers were marked this way. The MacRobert in question was Lady Rachel MacRobert of Douneside, Aberdeenshire, and her 'reply' was due to having lost all three sons in flying incidents: two during the war. Roderic was a Hurricane pilot killed while on an airfield attack in the Middle East and Iain was lost while searching the North Sea for downed airmen. Lady MacRobert wrote a cheque for £25,000 to 'buy a bomber to carry on their work in the most effective way'. She later sent a further £20,000 to buy four Hurricanes. Lady MacRobert also gave a country house as a leave centre for RAF personnel.

 BOMB GUNS FOR AEROPLANES

'For the purpose in making some tests in dropping bombs on to warships, at Garden City, New York, Mr Clifford B. Harmon fitted a novel "gun" to his Farman biplane. It consisted of a steel tube placed vertically, the lower end being closed by a hinged door. An ingenious arrangement of mirrors shows the operator exactly when the tube is pointing directly to the object he wishes to strike, and by merely pressing a button the door is released, and the bomb drops to its mark.'

(*Flight* magazine, 3 September 1910)

✈ BLACK BOX BOOKS

Ten of the books available on why aircraft crash:

1   *Why Planes Crash – An Accident Investigator Fights for Safe Skies*
2   *Black Box: Aircrash Detectives – Why Air Safety Is No Accident*
3   *The Black Box: All-New Cockpit Voice Recorder Accounts of In-Flight Accidents*
4   *Beyond the Black Box: The Forensics of Airplane Crashes*
5   *Air Disasters: Dramatic Black Box Flight Recordings*
6   *Air Disasters: The Truth Behind the Tragedies*
7   *Aviation Disasters: The World's Major Civil Airliner Crashes Since 1950*
8   *Military Aviation Disasters: Significant Losses Since 1908*

9   *Early Aviation Disasters: The World's Major Airliner Crashes Before 1950*
10   *Snakes in the Cockpit: Images of Military Aviation Disasters*

Black boxes are not black but painted in high-visibility markings, usually orange and white, to aid their location following an incident. Their name reputedly derives from one of the earliest flight recorders, which marked an image on scrolling photographic film, deviating according to changes in altitude, speed etc. The photographic process required absolute darkness inside the box, hence 'black', so the theory goes.

PHANTOM VS JAGUAR

The early summer of 1982 saw British aircraft in combat with their counterparts from Argentina. Far from the South Atlantic, another aerial encounter was to claim a 'kill', albeit an unintended one. An RAF Phantom from 92 Squadron, based in Germany, was scrambled as part of a tactical exercise. The fighter was carrying live weapons. During a combat air patrol two RAF Jaguars were detected and pursued. The Phantom slid in behind one of them and the pilot went through the procedures required to launch a Sidewinder heat-seeking missile. Pulling the trigger, the last thing the pilot expected to see was an actual missile leave his aircraft and streak towards the Jaguar up ahead. Within seconds the Sidewinder exploded causing catastrophic damage. Luckily the Jaguar pilot was quickly alerted to his situation by his flight leader and he ejected immediately, landing without injury.

An inquiry found that the Phantom's master armament switch should have had a piece of tape covering it, to prevent inadvertent operation. In the heat of the moment the pilot had gone through the drill forgetting his aircraft was carrying live missiles. The Phantom pilot and his navigator were both court-martialled.

BIRD STRIKES

The Civilian Aviation Authority collected figures in 2010 on which types of birds were most involved in bird strikes:

Gulls*	297	Swifts	81
Swallows	165	Wood pigeons	71
Skylarks	93	Pigeons	70

| Kestrels | 49 | Meadow pipits | 43 |
| Starlings | 49 | Rooks | 38 |

* Types included: Common Gull, Black-headed Gull, Herring Gull, Lesser Black-backed Gull, Great Black-backed Gull, Yellow-legged Gull, Great Black-headed Gull and 'Gull (unspecified)'.

MISSION ACCOMPLISHED

'We are safely on the other side of the pond. The job is finished.'

Lieutenant Commander Albert Read, US Navy, pilot of Curtiss NC-4 flying boat, after first aerial crossing of the Atlantic, May 1919.

FIRST SUPERSONIC EJECTION

The first successful supersonic ejection involving a live test subject took place in 1962. The test pilot chosen for this landmark event wasn't selected from an air force squadron, but from a zoo: it was a bear named Yogi. This specific species was chosen as a bear's internal body structure resembles that of a human's. The test, which took place at 35,000ft, was designed to prove the effectiveness of an enclosed escape capsule fitted in the American B-58 Hustler nuclear bomber. Yogi survived her forced ejection into the 870mph slipstream and subsequent eight-minute descent to the ground, although she was later subjected to an autopsy.

ENID

'Enid' is the name for part of the Red Arrows' formation. The world-renowned RAF Aerobatic Team fly nine Hawk trainer jets and while they fly formations that include all aircraft of the team at times, they often split into smaller components. The front five form Enid. When flying as a nine-ship the Arrows have flown these formations: Diamond Nine, Big Nine, Card, Tango, Big Vixen, Corkscrew, Goose, Apollo, Vixen, Nine Arrow, Eagle, Fred and Concorde.

✈ GREAT PLANES: DOUGLAS DC-3/DAKOTA

Designed as a civilian passenger aeroplane, the Douglas DC-3 first flew in 1935. It was originally known as the DST – the Douglas Sleeper Transport – equipped with fourteen sleeping berths and a honeymoon suite. It became a conventional airliner and was then adapted for military service. Almost 2,000 were used by the RAF to deliver supplies and drop parachutists on D-Day and at Arnhem. Over 10,000 DC-3s were built and some continue to be operated over eighty years after the type's first flight. The last scheduled passenger service to fly the DC-3 is operated in Canada by Buffalo Airways. Renowned British test pilot John Farley wrote: 'I consider it to have been the world's first truly great aeroplane.'

This Second World War-era Dakota is still in flying condition and is operated by the RAF's Battle of Britain Memorial Flight, based at RAF Coningsby in Lincolnshire. It is used for pilot familiarisation and also as a display aircraft. *Derek N. Ferguson*

✈ _HINDENBURG_

'It's practically standing still now. They've dropped ropes out of the nose of the ship, and they've been taken a hold of down on the field by a number of men ... The back motors of the ship are just holding it just, just enough to keep it from ... it burst into flames! It burst into flames, and it's falling, it's crashing! Watch it! Watch it, folks! Get out of the way! Get out of the way! Get this, Charlie! Get this, Charlie! It's fire – and it's crashing! It's crashing terrible! Oh my, get out of the way, please! It's burning and bursting into flames, and the ... and it's falling on the mooring mast and all the folks agree that this is terrible, this is the worst of the worst catastrophes in the world ... Oh, the humanity and all the passengers screaming around here ... I'm going to step inside where I cannot see it ... I'm gonna have to stop for a minute because I've lost my voice. This is the worst thing I've ever witnessed.'

Herbert Morrison, radio announcer, Lakehurst Naval Air Station, describing the LZ129 _Hindenburg_ accident, 6 May 1937.

✈ LOVELY AIRCRAFT NAMES

Armstrong Whitworth Siskin
Beardmore Inflexible
Blackburn Cubaroo
Boulton Paul Overstrand
Boulton Paul Sidestrand
Cierva Air Horse
Dart Flittermouse
Fairey Fawn
Fairey Fox
Fairey Gordon

Fairey Hamble Baby
Fleet Fawn
Granger Archaeopteryx
Hawker Hedgehog
Huff-Daland Dizzy Dog
Miles Whitney Straight
Parnall Heck
Percival Merganser
Republic Rainbow
Tipsy Nipper

✈ PASSENGER NUMBERS TAKE OFF

Passengers (in millions) passing through UK airports over a twenty-year period:

1990 – 101.7
1991 – 95.2
1992 – 105.5
1993 – 111.6

1994 – 121.5
1995 – 128.7
1996 – 135.2
1997 – 145.9

British Airways airliners including long-haul Boeing 747s and Boeing 777s can be seen from a departing aircraft at London's Heathrow Airport. *Tony Hisgett*

1998 – 158.1	2004 – 214.9
1999 – 167.6	2005 – 227.4
2000 – 179.1	2006 – 234.4
2001 – 180.5	2007 – 239.9
2002 – 188.0	2008 – 235.3
2003 – 199.2	2009 – 218.1

✈ NOTABLE DEATHS IN AIRCRAFT CRASHES

Torino FC (1949)
Manchester United FC (1958)
Aaliyah (musician)
Patsy Cline (musician)
Hansie Cronje (cricketer)
John Denver (musician)
Stevie Gaines (musician, Lynyrd Skynyrd)
Bill Graham (music promoter)
Dag Hammarskjöld (UN Secretary General)
Graham Hill (racing driver)
Steve Hislop (motorcycle racer)
Buddy Holly (musician)
Leslie Howard (actor)
John F. Kennedy Jr (son of the President)
Carole Lombard (actress)
Rocky Marciano (boxer)
Colin McRae (rally driver)
Glenn Miller (musician)
Vic Morrow (actor)
Audie Murphy (actor)
Ricky Nelson (musician)
Otis Redding (musician)
Jim Reeves (musician)
Randy Rhoads (musician)
J.P. Richardson (musician, The Big Bopper)

Will Rogers (actor)
Payne Stewart (golfer)
Ritchie Valens (musician)

Ronnie Van Zant (musician, Lynyrd Skynyrd)
Stevie Ray Vaughan (musician)

'DICTA BOELCKE' – THE RULES OF DOGFIGHTING

Oswald Boelcke was one of the leading fighter pilots of the First World War. He was the mentor of another German ace, Manfred von Richthofen – the 'Red Baron'. Boelcke, who shot down forty aircraft, issued his rules of dogfighting:

1 Ensure positional advantage; keep the sun behind you to make it hard for the opponent to see you.
2 Once an attack is begun, it must always be carried through.
3 Only open fire when the target is fully in your sights.
4 Pay careful attention to the manoeuvres of the enemy, do not be tricked, follow him to the ground to make sure he is not faking his end.
5 Attack only from behind.
6 If you are attacked you should not run, but turn to face your opponent.
7 Over enemy territory, always have your return route in mind.
8 The squadron should attack together in small groups, but no two aircraft should attack the same target.

Boelcke was killed when he fell foul of one of his own rules and collided with another German aircraft when engaging a target.

DID YOU KNOW?

29.44 million:
Number of delayed, damaged or stolen passenger baggage items each year, worldwide.

A STRIKING DISPLAY

Pageants, or airshows as we now describe them, took place at Hendon, now the site of the RAF Museum. Large crowds would gather to see the dashing young men in their flying machines. This report appeared in *The Times*, in July 1923:

'The Royal Air Force Pageant, held on Saturday at Hendon Aerodrome as usual, had a special importance this year for apart from the visit of the King and Queen, there was combined with the spectacular display of flying a review of some of the latest and most formidable aircraft from which the RAF is being equipped.

'To the general public one aeroplane is very similar to another, but these latest types embody some important developments. Their speed (over 150mph, 2 miles up in the air in some cases), rapid climb, and power of manoeuvring, as well as the highest degree of skill shown by the pilots, must have impressed the many foreign delegates present, as well as the very large gathering of the general public. Probably eighty thousand people witnessed the Pageant from the aerodrome itself, in addition to the many thousands who saw it from points outside the ground. The motor-car enclosures and the private boxes on the rails of the course almost resembled Ascot in the smartness of the frocks worn by the women.

'Some nearly perfect formation flying by No 39 Squadron in de Havilland 9a's resembled ground drill transferred to the air, and the low bombing flights of five Sopwith Snipes were spectacular, if not particularly accurate in their results. Flight Lieutenant Longton's crazy flying a few feet from the ground demonstrated humorously most of the things one does not do in the air, and five Snipes showed effectively how to break up and harry a trio of heavy bombers. There were, in addition, several interesting races.'

AIR LINES

A few memorable advertising slogans and straplines:

You and I were meant to fly.
– Air Canada

Have you ever done it the French way?
– Air France

Our passengers get the best of everything.
– American Airlines

Does your wife know you're flying with us?
– Braniff

The world's favourite airline.
— British Airways

We never forget you have a choice.
— British Caledonian

We really move our tail for you.
— Continental

The airline with the big jets.
— Delta Airlines

The web's favourite airline.
— easyJet

There's no better way to fly.
— Lufthansa

I'm Cheryl. Fly me.
— National Airlines

Catch our smile.
— Pacific Southwest Airlines

Live today. Tomorrow will cost more.
— Pan American

Enjoy our company.
— Sabena

Singapore Girl, you're a great way to fly.
— Singapore Airlines

Remember what it was like before Southwest Airlines?
— Southwest Airlines

Our sign is a promise.
— Swissair

Fly the finest ... Fly TWA.
— Trans World Airlines

It's time to fly.
– United

You never forget your first time.
– Virgin Atlantic

INCREDIBLE MACHINES: XB-70 VALKYRIE

The North American XB-70 Valkyrie was intended to be a long-range, high-speed strategic bomber. It was an impressive beast. The Valkyrie had six afterburning turbojet engines. It could fly at three times the speed of sound, and was designed to cruise at 70,000ft, well away from any intercepting fighters. It was almost 190ft long and had a wingspan of over 100ft. The aircraft was made largely of stainless steel. Its wing formed a huge delta shape, with a rotatable outer part to help stability and generate more lift. Only two were built, and one was lost due to a collision with a chase aircraft. The remaining aircraft was retired three years later after serving as a high-speed research vehicle.

The futuristic shape of the North American XB-70 Valkyrie as it takes off from Edwards Air Force Base, California. The canard control surfaces can be seen clearly on either side of the aircraft's thin forward fuselage. *NASA*

✈ BABIES ON A PLANE

The first recorded baby to be born in flight came into the world 6,000ft above it. In the summer of 1922 the baby's mother, Madame Georges Breyer of Lyons, was in the south of Italy when she felt the birth was becoming imminent. She chartered a plane to take her to Naples but was 40 miles short when the baby boy arrived. The infant was named Guynemer, after the French First World War pilot. It was reported the mother purchased the aircraft as a souvenir.

Another air-born arrival was planned deliberately by the parents. In October 1929 in Florida, two parents-to-be chartered a Fokker Tri-motor. The plane was directed to fly around at 1,200ft over a courthouse, until the baby was duly delivered. The name the baby was given? Airlene.

✈ PAYNE STEWART

Payne Stewart was a well-loved professional golfer, noted for his outlandish outfits. On 25 October 1999, Stewart and his three companions boarded a private Learjet to fly from Orlando in Florida to Dallas, Texas. When the aircraft started to drift off course at high altitude, air traffic controllers requested an aerial inspection. A US Air Force F-16 was duly vectored towards the private jet. The pilot drew alongside the Learjet at 45,000ft and noticed that the cockpit windows were either frosted or covered in condensation – a tell-tale sign that the occupants were probably already dead. The Learjet flew on, controlled by its autopilot, until it ran out of fuel almost four hours after take-off. It crashed in a field in South Dakota. The probable cause was a loss of pressure in the cabin leading to the crew and passengers becoming incapacitated. The lack of oxygen at such high altitudes would have caused them slowly to lose consciousness, drifting off to their deaths.

✕ DID YOU KNOW?

3,500,000:
The number of people who attended airshows in the UK in 2006.

WORDS OF WISDOM

'A good landing is one you can walk away from.'

Anon.

NICKNAMES

A selection of the more unusual or noteworthy nicknames of post-war British test pilots:

Roland P. 'Bee' Beaumont
Eric 'Winkle' Brown
Frederick J. 'Jeep' Cable
Digby 'Digger' Cotes-Preedy
John 'Cats Eyes' Cunningham
R.J. 'Roly' Falk
Eric 'Jumbo' Genders
A.E. 'Ben' Gunn
W.T. 'Doddy' Hay

'Chunky' Horne
C.T.D. 'Sox' Hosegood
'Pee Wee' Judge
'Pingo' Lester
G.R.I. 'Sailor' Parker
A.J. 'Bill' Pegg
E.J. 'Jimpy' Shaw
Joseph J. 'Mutt' Summers
Trevor S. 'Wimpy' Wade

BAND NAMES

1 Led Zeppelin (German airships)
2 Spitfire (British fighter of the Second World War)
3 U-2 (Lockheed spy plane)
4 The B-52s (Cold War bomber)
5 Bell X-1 (First aircraft to break the sound barrier)
6 Blue Aeroplanes (Self-explanatory)
7 Foo Fighters (Early name given to UFOs seen by military pilots)
8 El Ten Eleven (Lockheed L-1011 airliner)
9 SR-71 (Lockheed ultra-fast spy plane)
10 Sopwith Camel (First World War fighter)

In 1974 Robert Calvert, ex-singer with Hawkwind, released a concept album called *Captain Lockheed and the Starfighters* which took as its theme the sale of F-104 Starfighters to Germany. To date it is the only known album dedicated to an international fighter sale.

✈ DID YOU KNOW?

487:

The world record number of aircraft types flown by legendary British test pilot Captain Eric 'Winkle' Brown. Born in 1919, Brown flew in the Fleet Air Arm in the Second World War and went on to fly various captured German aircraft, as well as the early British jets. He was the first to land a jet aircraft on an aircraft carrier and holds the record for the largest number of deck landings.

✈ GREAT PLANES: MOSQUITO

The nickname 'The Wooden Wonder' conveys how the Mosquito was viewed by its crews. The de Havilland-built 'Mossie' served in a variety of different roles: as an unarmed bomber, a low-level anti-shipping strike aircraft, a night-fighter, a fighter-bomber, and a photo-reconnaissance aircraft. It could fly unarmed because its top speed was fast enough to outrun opponents. Part of the reason behind the Mosquito's superb performance was the power of its engines – it had two Rolls-Royce

The Mosquito proved to be a popular subject for plastic kit makers and this artwork by Roy Cross is for the Airfix 1:72 scale model. It shows a Mark XVIII equipped with a 57mm Molins gun used in the anti-shipping role. *Hornby Hobbies Ltd*

Merlins. During its initial flight tests, it was able to outpace Spitfires. The Mosquito had the lowest operational loss rate of any Allied aircraft during the Second World War, and for long-range operations, was able to carry the same bomb load as the four-engined American B-17 Flying Fortress.

Amongst its many accomplishments, the Mosquito took part in two famous low-level precision bombing raids on the Amiens prison, housing French Resistance fighters, and the Gestapo headquarters in Copenhagen. Another mission was planned to interrupt a speech by the head of the German Luftwaffe, Herman Goering. The raid took place during the day to that feared destination, Berlin, and was achieved successfully with the sounds of the bombs exploding appearing in the broadcast. Goering said: 'It makes me furious when I see the Mosquito. I turn green and yellow with jealousy.' Despite over 7,700 Mosquitos being built, there are currently none flying in Britain.

DOORS TO MANUAL

The first female flight attendants took to the air in 1930 for Boeing Air Transport (later United Airlines). These fledgling 'sky girls' were given specific instructions through a stewardess manual written by the very first female flight attendant, Ellen Church, and Boeing district manager Steve Stimpson. These are some of the instructions:

♦ Remember at all times to retain the respectful reserve of a well-trained servant.

♦ Captains and cockpit crew will be treated with strict formality while in uniform. A rigid military salute will be rendered to the captain and co-pilot as they go aboard.

♦ Use a small broom on the floor prior to very flight.

♦ Check the floor bolts on wicker seats to ensure they are securely fastened down.

♦ Swat flies in cabin after take-off.

♦ Warn passengers against throwing lighted smoking butts or other objects out of the windows, particularly over populated areas.

♦ Carry a railroad timetable in case the plane is grounded somewhere. Stewardesses are expected to accompany stranded passengers to the railroad station.

♦ Keep an eye on passengers when they go to the lavatory to be sure they don't mistakenly go out the emergency exit.

✈ WORDS OF WISDOM

'I am not allowed to say how many planes joined the raid, but I counted them all out and I counted them all back. Their pilots were unhurt, cheerful and jubilant, giving thumbs-up signs.'

BBC reporter Brian Hanrahan, HMS *Hermes*, Falkland Islands, May 1982.

✈ JAMES BOND AIRCRAFT

Some of the aircraft that have appeared in James Bond films:

Aero L-39 Albatros
Tomorrow Never Dies
Aérospatiale HH-65A Dolphin
Licence to Kill
Airship Industries Skyship 500
A View to a Kill
Antonov An-124
Die Another Day
Avro Vulcan
Thunderball
Bede BD-5J Micro-jet
Octopussy
Beechcraft Model 18
Octopussy
Bell JetRanger
The Spy Who Loved Me
Boeing B-17G
Thunderball
Douglas DC-3
Quantum of Solace

Eurocopter AS355 Ecureuil 2
Tomorrow Never Dies
Eurocopter Tiger
GoldenEye
Handley Page Jetstream
Moonraker
Hiller UH-12
From Russia with Love
Kawasaki KV-107 II
You Only Live Twice
Lockheed JetStar
Goldfinger
Pilatus PC-6
GoldenEye
Piper PA-28 Cherokee
Goldfinger
Republic RC-3 Seabee
The Man with the Golden Gun
Wallis WA-116 Autogyro 'Little Nellie'
You Only Live Twice

 F-4 PHANTOM SPIN RECOVERY TECHNIQUE

Upright Spin Recovery:

1   Positively determine spin direction.
2   Maintain full forward stick and neutral rudder, and supply full aileron in the direction of the spin. (Right turn needle deflection, right spin, right ailerons.)
3   When the aircraft unloads (negative G) and/or yaw rate stops, neutralise the ailerons and fly out of the unusual attitude.
4   Do not exceed 19 units (angle of attack) during dive pull out.
5   If still out of control by 10,000ft above the terrain – eject.

Flat Spin Recovery:

There is no known technique for recovery from a flat spin.

✈ APOLLO FLYERS

The astronauts who flew as part of the Apollo programme are famous for their exploits in space, especially those who journeyed to the moon. Before they joined NASA all but one of those assigned to Apollo missions had flown in one of the United States' military air arms:

US Air Force

Edwin E. 'Buzz' Aldrin* (LMP Apollo 11)
William A. Anders (LMP Apollo 8)
Frank F. Borman (CDR Apollo 8)
Michael Collins (CMP Apollo 11)
Charles M. Duke Jr* (LMP Apollo 16)
Donn F. Eisele (CMP Apollo 7)
Virgil I. 'Gus' Grissom (CDR Apollo 1)
Fred W. Haise (LMP Apollo 13)
James B. Irwin* (LMP Apollo 15)
James A. McDivitt (CDR Apollo 9)
Stuart R. Roosa (CMP Apollo 14)

David R. Scott* (Apollo 9, CDR Apollo 15)
Russell L. 'Rusty' Schweikart (LMP Apollo 9)
Donald K. 'Deke' Slayton (DMP Apollo-Soyuz Test Project)
Thomas P. Stafford (CDR Apollo 10, CDR Apollo-Soyuz Test Project)
John L. 'Jack' Swigert (CMP Apollo 13)
Edward H. White II (Senior Pilot Apollo 1)
Alfred M. Worden (CMP Apollo 15)

US Marine Corps

Vance D. Brand (CMP Apollo-Soyuz Test Project)

Ronnie W. 'Walter' Cunningham (LMP Apollo 7)

US Navy

Neil A. Armstrong* (CDR Apollo 11)
Alan L. Bean* (LMP Apollo 12)
Eugene A. 'Gene' Cernan* (LMP Apollo 10, CDR Apollo 17)
Roger B. Chaffee (Pilot Apollo 1)
Charles 'Pete' Conrad* (CDR Apollo 12)
Ronald E. Evans (CMP Apollo 17)
Richard F. Gordon (CMP Apollo 12)

James A. Lovell (CMP Apollo 8, CDR Apollo 13)
Thomas Kenneth Mattingly II (CMP Apollo 16)
Edgar D. Mitchell* (LMP Apollo 14)
Walter M. 'Wally' Schirra (CDR Apollo 7)
Alan B. Shepard* (CDR Apollo 14)
John W. Young* (CMP Apollo 10, CDR Apollo 16)

Civilian

Harrison H. 'Jack' Schmitt (LMP Apollo 17)

* – Walked on the moon
CDR – Commander
CMP – Command Module Pilot
DMP – Docking Module Pilot
LMP – Lunar Module Pilot

✈ TARGET INDICATOR BOMB COMPONENTS

A list of all the parts of No 1, Mark 1, Target Indicator Bomb (Second World War):

Tail Plate, Bayonet Slot, Packing Disc, Suspension Lug, Steel Tube, Felt Strip, Non-Delay Candle, Primed Cambric Tube, Notched Washer, Steel Nose, Leather Washer, Transit Plug, Gunpowder, Burster Container, Cotton Cambric Bag, Flash Hole, Ejector Plate, Steel Central Tube, Wooden Batten, Division Disc, Primed Cambric Tube, Rivet.

 EXTREME AIRPORTS

Saba (Juancho E Yrausquin Airport)

Possibly the most precarious runway in the world. The small Caribbean island of Saba has no beaches, only cliffs that drop straight into the sea and the island's only runway is built on a very small promontory at one end. Not only is the runway in a hazardous location, it is also short (1,312ft), making it the world's smallest commercial runway. Unlike most other airports, there is no run-off area, just a 200ft drop to the ocean.

St Maarten (Princess Juliana International Airport)

Most visitors who fly to Saba reach it via the much bigger airport at St Maarten, another of the extreme airports of the world. It has mountains at one end and the sea at the other. Although suitable for large airliners such as Boeing 747s, the 7,500ft runway doesn't leave much room for error so aircraft aim to land as close to the start of the runway as possible. There is much fun to be had for those brave enough to stand at the fence experiencing the thrust of departing jets.

A US Airways Boeing 757 airliner on final approach to St Maarten Airport shows how near the runway is to the beach. *Lawrence Lansing*

Courchevel

Courchevel is a ski resort in the French Alps. With the Alps not renowned for their flat open spaces, it should come as no surprise to discover the town's runway is built on a slope: an 18.5 per cent slope. To add to the mix the runway is only 1,722ft long and is situated at an elevation of over 6,500ft. Take-offs are made down the hill and landings up the hill so it is important for pilots to ensure the runway is for their sole use when about to land or take off. The approach and departure routes are made through the mountain range and it is possible for circuits to be flown below the level of the control tower. The opening sequence of the James Bond film *Tomorrow Never Dies* was filmed here.

Lukla (Tenzing-Hillary Airport)

Lukla is another mountainside airport with a heavily sloped runway. It is thirty-five minutes' flying time from Kathmandu in north-east Nepal and takes trekkers headed to and from Mount Everest, saving them a five-day journey by foot. The runway slopes at a (slightly) more benign angle than Courchevel of 11.5 per cent. It is shorter, at 1,600ft, and sits at an elevation of 9,380ft, not far from the altitude where pilots and passengers would require oxygen. Unsurprisingly there have been fatal accidents including one in 2008 when eighteen were killed on a landing approach when the aircraft hit the ground short of the runway.

Barra

Barra in the Outer Hebrides has a unique claim. It is the only airport set on a beach that operates scheduled flights. When the tide comes in the runways disappear under sea water. Three runways are marked out in the sand to ensure aircraft can make an approach and landing into the wind. Around 10,000 passengers make the journey each year. Visitors to the beach are advised to stay clear when aircraft are operating.

✈ FIRST PASSENGER FATALITY

The first death of a passenger in a heavier-than-air aircraft was the unfortunate Lieutenant William F. Selfridge of the US Army. He was being taken on a demonstration flight by Orville Wright when the aircraft got into difficulties and crashed. Selfridge was fatally injured and Wright sustained serious injuries. At one point the Wright Flyer was heading towards Arlington National Cemetery before being steered away. Lieutenant Selfridge was later buried there. An American newspaper described the crash:

'While the machine was encircling the drill grounds a propeller blade snapped off and hitting some other part of the intricate mechanism caused it to overturn in the air and fall to the ground, enveloping the two occupants in the debris. Soldiers and spectators ran across the field to where the aeroplane had fallen and assisted in lifting Mr Wright and Lieutenant Selfridge from under the tangled mass of machinery, rods, wires and shreds of muslin. Mr Wright was conscious and said, "Oh hurry and lift the motor." Lieutenant Selfridge was unconscious. His head was covered with blood and he was choking when the soldiers extricated him from under the machine.'

The Daily Republican, 18 September 1908.

✈ USES FOR DISUSED AIRFIELDS IN BRITAIN

Army barracks
Business parks
Car plants
Car storage sites
Car testing tracks
Caravan sites
Docks
Farmland
Golf courses
Housing estates
Industrial estates

Motorsport race tracks
Motorways
Museums
Nuclear power stations
Prisons
Quarries
Racecourses
Schools
University grounds
Zoos

 MODEL AIRCRAFT KIT MANUFACTURERS

Academy Models
Accurate Miniatures
Aeroplast
Airfix
Alanger
Aoshima
Ark Models
AZ Model
Bilek
Dragon

Eastern Express
Eduard
ESCI
Fine Molds
Frog
Fujimi
Hasegawa
Heller
Hobbyboss
Hobbycraft

Huma
ICM
Italeri
Kangnam
Kinetic
Legato
Mach2
Maquette
Matchbox
Mini Hobby Models
Mirage Hobby
Modelcraft
Monogram
MPM

Pavla Models
Pegasus Hobbies
PM Model
Revell
Roden
SMER
Special Hobby
Tamiya
Testors
Trumpeter
UM
Valom
Xtrakit
Zvezda

✈ SELECTED BRITISH CIVILIAN REGISTRATIONS

G-ALAN, G-AMEN, G-ARAB, G-BABE, G-BALD, G-BANE, G-BARK, G-BEAK,
G-BEAR, G-BEER, G-BELT, G-BERT, G-BILL, G-BOKE, G-BOMB, G-BORE,
G-BUFF, G-BUMF, G-BUNT, G-BUTT, G-CAMP, G-CHOP, G-CLAW, G-COOL,
G-COPS, G-CULL, G-DAFT, G-DENT, G-DING, G-DIRE, G-DIVE, G-DOGS,
G-DUNG, G-FIRE, G-FIRM, G-FLEW, G-FLIP, G-FRED, G-FREE, G-FROG,
G-FUEL, G-GASP, G-GLAD, G-GLUG, G-GOAL, G-GONE, G-GOOD,
G-GORE, G-GOSH, G-GRIN, G-GUNS, G-GUST, G-HANS, G-HAUL,
G-HAWK, G-HEAT, G-HELP, G-HENS, G-HERO, G-HERS, G-HEWS, G-HIGH,
G-HILL, G-HIRE, G-HOBO, G-HOGS, G-HOLE, G-HOLY, G-HOME, G-HONE,
G-HONK, G-HOOK, G-HOOP, G-HOPE, G-HORN, G-HUFF, G-HUGE,
G-HUGS, G-HUMP, G-HUNK, G-HUNT, G-IAIN, G-IANT, G-IDDY, G-IJOE,
G-IRLS, G-IGLE, G-IVEN, G-JAZZ, G-JEST, G-JETS, G-JOCK, G-JOKE, G-JUMP,
G-KILT, G-KINK, G-KNOB, G-KING, G-LACE, G-LADS, G-LARE, G-LASH,
G-LASS, G-LAZE, G-LEAP, G-LEGS, G-LENS, G-LICK, G-LIDS, G-LIFE, G-LIFT,
G-LILY, G-LIME, G-LINE, G-LISA, G-LIVE, G-LOBE, G-LOCH, G-LOOK,
G-LOOP, G-LORY, G-LOSS, G-LOST, G-LOVE, G-LUCK, G-LUCY, G-LUKE,
G-LULU, G-LUST, G-MACH, G-MALE, G-MALL, G-MALT, G-MEAT, G-MELT,
G-MOON, G-MOTH, G-NOSE, G-NUTS, G-OLLY, G-OOSE, G-PAWS,
G-PEST, G-PETE, G-PIES, G-PIGS, G-PING, G-PINT, G-PIPE, G-PNUT, G-POKE,
G-POPE, G-PORK, G-PORT, G-POSH, G-POST, G-PROP, G-PUBS, G-PUFF,
G-PULL, G-PUMP, G-PUNK, G-PURE, G-PURR, G-PUSH, G-PUSS, G-RACE,
G-RAFT, G-RAVE, G-REED, G-REAT, G-RIBS, G-RIPS, G-RISE, G-ROIN,
G-ROLL, G-ROPE, G-ROWL, G-ROWS, G-RUFF, G-RUMP, G-RUNT, G-SAFE,

G-SAGE, G-SCOT, G-SHOT, G-SHOW, G-SLAM, G-SLUG, G-SNIP, G-SOAR, G-SOFT, G-SOLO, G-SOUP, G-SPIT, G-SPOT, G-SWOT, G-TAIL, G-TART, G-TINS, G-TIRE, G-TOAD, G-TOBY, G-TOFF, G-TOLL, G-TONE, G-TOOL, G-TOWS, G-TOYS, G-UEST, G-UIDE, G-UILD, G-USTO, G-USTS, G-ULPS, G-UILT, G-VIRG, G-WASP, G-WEED, G-WEEK, G-WELL, G-WENT, G-WHAM, G-WHAT, G-WIFE, G-WIND, G-WING, G-WINK, G-WINS, G-YOGI, G-YELL, G-ZOOM, G-ZORO, G-ZULU.

 DID YOU KNOW?

12:
Number of .303 Browning machine guns fitted on the Hawker Hurricane IIB.

 SPECIAL BALLOONS

Ballooning is one of the oldest forms of flying and it continues to the present day. Calm summer mornings and tranquil evening skies around the world see the rising shapes of balloons carrying passengers on pleasure flights. Many balloons are now also used for promotional purposes, and varied and interesting shapes have sprung up as a result, including:

This specially shaped hot-air balloon was constructed to resemble a panda and is a relatively simple conversion from the normal designs it is displayed beside. *Jim Bahn*

Beer can
Birthday cake
Bulldog
Bunch of tulips
Condom
Darth Vader
Disney Magic Kingdom
FA Cup
Fire extinguisher
Flying dragon
Golliwog
Halloween pumpkin
Ice cream cone
Jesus
Lightbulb
Mineral water bottle
Motorbike
Mouse
Newspaper
Oil can
Orient Express
Pair of jeans
Panda
Parachutist
Parrot
Polar bear
Santa Claus
Scarecrow
Scotsman
Shopping trolley
Snowman
Space shuttle
Spark plug
Sports car
Stagecoach
Stork carrying a baby
Sun
Thatched cottage
Tobacco pipe
Toilet rolls
Training shoe
UFO
Vincent Van Gogh's head
Whisky bottle

✈ ON TOP OF THE WORLD

On 3 April 1933 two aircraft made the first flights over the summit of Mount Everest. The two biplanes were a Houston-Westland (named in honour of Lady Houston who sponsored the expedition) which was crewed by Colonel L.V.S. Blacker and Lord Clydesdale (14th Duke of Hamilton and the man Rudolf Hess flew to meet in 1941), and a Westland Wallace flown by Flight Lieutenant D.F. McIntyre and S.R. Bonnett, a cameraman for Gaumont-British Film Corporation. Lord Clydesdale's report appeared in *The Times*:

'On approaching Lhotse, the southern peak of the Everest group, the ground rises at a steep gradient, and both machines experienced a steady down current due to deflection of the west wind over the mountain, causing a loss of altitude of 1,500 feet, despite all our efforts to climb. Both aeroplanes flew over the summit of Everest at 10.05, clearing it by 100ft near the summit, but no bumps were felt by either aircraft. Fifteen minutes were spent flying in the neighbourhood of the summit ... The visibility of distant high peaks was very good. The great Himalaya range could be seen extending to great distances and provided a magnificent spectacle.'

Both aircraft landed safely, although Bonnett had fallen unconscious after his oxygen supply was interrupted (the aircraft had open cockpits). A second, unauthorised flight was made in order to secure adequate photographs a few weeks later.

✈ KEYBOARD REFERENCE GUIDE

Keyboard commands for *F/A-18 South Korea* computer game (PC version):

Keypad 4	Aileron Left	Shift S	Service
Keypad 6	Aileron Right	Shift E	Eject
Keypad 8	Elevator Down	D	Damage Display
Keypad 5	Elevator Up	E	Engine Display
,	Rudder Left	S	Stores Display
/	Rudder Centre	[	Cycle Air-to-Air
.	Rudder Right		Weapons
=	Increase Thrust	]	Cycle Air-to-Ground
-	Decrease Thrust		Weapons
Backspace or Delete Afterburner		Shift]	CCIP/Auto toggle
Space	Speed Brake	J	Jettison Station Select
G	Gear	'	Release Flare
F	Flaps	;	Release Chaff
H	Hook	C	ECM Toggle
Shift D	Dump Fuel	Enter	Designate/Release
Shift F	Refuel		weapons

✈ ESSENTIAL AIRSHOW EQUIPMENT

Camera
Telephoto lenses
Wide-angle lenses
Spare memory cards
Spare batteries
Lens cloths
Filters
Lens hoods
Spare camera body

Binoculars
Walking boots
Jacket (rainproof)
Trousers (rainproof)
Sunglasses
Sunscreen
Hat
Money
Rucksack

Fold-up seat   Handkerchief
Bottled water   Antibacterial hand gel
Flask   Headache pills
Antihistamines   Tartan rug

FEMALE FIRST

On 15 May 1930 Ellen Church became the world's first female flight attendant. She worked on a Boeing Model 80A on the route from San Francisco to Cheyenne in Wyoming for Boeing Air Transport. Church had suggested that nurses would be useful and comforting for passengers nervous of flying. Her suggestion was taken forward by a Boeing manager, Steve Stimpson. However, when he first contacted his head office with the suggestion of employing stewardesses on board aircraft he received a one-word reply: 'No.' This was not the end and eventually the idea was given a trial period of three months. Eight 'sky girls' were initially hired. It was a condition of employment that they were registered nurses and also single, although one was in fact married. By 2008 there were 98,700 flight attendants in employment in the United States.

STREET NAMES OF HATFIELD

Hatfield was the location for de Havilland's factory where many of the manufacturer's most famous aircraft were built. A number of the town's streets were named appropriately:

Albatross Way   Gypsy Moth Avenue
Aviation Avenue   Horsa Gardens
Chipmunk Chase   Jetliner Way
Comet Road   Mosquito Way
Comet Way   Nimrod Drive
De Havilland Close   Oxford Place
Devon Mead   Queen Bee Court
Dragon Road   The Runway
Flamingo Close   Tiger Moth Way

✈ WHEN CONCORDE TRAVELLED BY ROAD AND SEA

In 2003 it was announced that Concorde would be retired. Many bids were made to British Airways to receive the airframes and one of the locations chosen was the Museum of Flight at the ex-RAF airfield at East Fortune in East Lothian. There were a number of issues in getting the Concorde selected, G-BOAA, to the museum from Heathrow. G-BOAA was not in an airworthy condition and even if it was, East Fortune's runway wasn't up to receiving such a large aeroplane. Concordes are too big to be airlifted so an ingenious solution had to be found. The wings, engines and tailfin were removed and the fuselage was mounted on to a specialised 48-wheel transporter. This was then steered through the streets from Heathrow in the middle of the night to the river Thames, where it was loaded on to a sea-going barge. The barge sailed up to Torness nuclear power station near Dunbar, where it was unloaded. The once supersonic airliner was then wheeled along the A1 at walking pace. (The transporter was steered by an operator on foot who used a remote-control steering box.) The last part of its journey required a special track to be laid across farm fields by the army. Once at the museum, all Concorde's pieces were reattached and it opened to the public in 2005. This finished one of the slowest journeys a Concorde ever made, and certainly one of the most unusual.

✈ FICTIONAL AIRCRAFT

Airwolf	A super-fast stealthy helicopter as seen in the US TV show of the same name.
Blue Thunder	Another highly capable helicopter, seen in the film and TV show of the same name.
MiG-31 Firefox	A Firefox is stolen by US pilot Mitchell Gant in Craig Thomas' novel *Firefox* and the subsequent film adaptation. The Russians did eventually build a MiG-31 but it had nothing like the capability of the fictional version. The real MiG was given the NATO codename 'Foxhound'.
MiG-28	These black fighter aircraft are from an unspecified country, taken on by US Navy aircrews in the film *Top Gun*. They bear a very strong resemblance to American Northrop F-5 aircraft.

Angel Interceptor	From the Gerry Anderson TV show *Captain Scarlet* the Interceptors were flown solely by women pilots (the 'angels') from a flying aircraft carrier called Cloudbase.
Blohm & Voss BV 38	This flying wing craft doesn't get off the ground in the Indiana Jones film *Raiders of the Lost Ark*. It is destroyed by the whip-carrying archaeologist before it can take off.
B-52 Megafortress	A highly modified version of the Cold War bomber, from Dale Brown's series of techno-thriller novels.
Thunderbird 2	This transporter aircraft is flown by Virgil Tracy and can carry a wide range of equipment and vehicles in interchangeable pods. Seen in Gerry Anderson's *Thunderbirds* TV programme.
Fireflash	A hypersonic airliner from *Thunderbirds*. Powered by nuclear engines to a top speed of Mach 6.
F/A-37 Talon	An advanced stealth fighter, the Talon operates off the aircraft carrier USS *Abraham Lincoln*. Appropriately enough, it is seen in the movie *Stealth*.

The F/A-37 Talon from the film *Stealth*. Images of this movie prop on the deck of a real US Navy aircraft carrier provoked speculation that it was an actual aircraft. *US Government*

✈ LUCKY ESCAPES: NO PARACHUTE

Flight Sergeant Nicholas Alkemade was a rear gunner on Avro Lancasters. In March 1944 his aircraft was heading to Berlin when it was attacked by German night fighters. The bomber was set on fire. Alkemade attempted to retrieve his parachute – the rear gunner's position was too small to accommodate the gunner and his parachute, so it was stored in the rear fuselage behind him – when he discovered it was alight. Rather than stay in the burning aeroplane he decided to jump. He fell 18,000ft and miraculously his descent was slowed by pine tree branches and then a snowdrift. His only injury was a twisted ankle. The Germans suspected him of being a spy until his burnt parachute was found in the plane's wreckage. He was made a prisoner of war and died in 1987.

✈ RAF SQUADRON NICKNAMES

2	Shiny Two
6	The Flying Can Openers
13	The Stabbed Cats
43	The Fighting Cocks
45	The Flying Camels
56	The Firebirds
111	The Tremblers
617	The Dam Busters

✈ AVIATION MATTERS

'The Women's Freedom League yesterday engaged Mr Herbert Spencer's airship, and Miss Muriel Matters, one of the leading members of the league, travelled some distance with Mr Spencer from Hendon, the object being to reach the Houses of Parliament. The airship started from the grounds of the Old Welsh Harp. On one side of the balloon in large black letters were the words "Votes for Women" and on the other side "Women's Freedom League."

'The wind prevented the balloon from following the route of the Royal Procession, and it was carried to Coulsdon, Surrey. The airship had been pursued from Hendon by Mrs How Martyn, Miss Elsie Craig, and others in a motor-car. Miss Matters stated that the balloon reached a height of 3,500ft, and passed the Houses of Parliament when they were so high that they could just distinguish the buildings. She did not think anyone looking from the House could have seen them. She threw out 56lb of hand-bills and these enabled the party in the motor-car to track them.'

(*The Times*, 17 February 1909)

MISSILES OF THE 1950S

Surface-to-air, air-to-surface or air-to-air missiles:

Armstrong Whitworth Canard
Bell Rascal
Boeing BOMARC
Bristol Bloodhound
De Havilland Firestreak
Convair Terrier
Douglas Nike
Douglas Skybolt

Fairey Fireflash
General Dynamics Tartar
Hawker Siddeley Red Top
Hughes Falcon
McDonnell Talos
Martin Bold Orion
Minneapolis-Honeywell Wagtail
North American Hound Dog

JANE HARRISON

Jane Harrison is the only woman to be awarded the George Cross in peacetime. On 8 April 1968 she was a flight attendant on board BOAC Boeing 707 registration G-ARWE when it suffered an engine fire on take-off. The airliner managed to turn back to Heathrow Airport and carried out a successful emergency landing. Evacuation procedures were put into place as the fire spread. Jane Harrison remained on board guiding passengers out of the emergency exits. Her body was found next to that of a disabled passenger who it is believed she had stayed behind to assist. She was 22 years old.

DID YOU KNOW?

519:

The number of Russian tanks destroyed by Luftwaffe pilot Hans Ulrich Rudel. Rudel flew the Junker Ju-87G Stuka 'PanzerJager' (tank-hunter). Its armament was two long-barrelled 37mm cannons. As well as this number of tanks, Rudel also destroyed numerous other targets including four Russian navy ships. He was Germany's most decorated airman, receiving the country's highest award: the Knight's Cross of the Iron Cross with Golden Oak Leaves, Swords and Diamonds. During a mission he received injuries to his right leg that necessitated amputation, although this did not prevent him carrying on combat flying. In May 1945 he flew west to escape being captured by the Russians and spent time as a prisoner of war. He died in 1982.

✈ CORONATION REVIEW

In July 1953 the RAF organised a review for the newly crowned Queen Elizabeth II. Odiham was chosen as the venue for a show of the air force's strength. The static display consisted of ground equipment and 318 aircraft which took the Queen an hour to drive round. An incredible 641 aircraft took part in the flypast, arriving from over forty different airfields. The half-hour-long flypast was rounded off by examples of future types for the RAF, including the three V-bombers:

Sycamore (1), Chipmunks (32), Prentices (12), Harvards (12), Oxfords (12), Ansons (12), Balliols (12), Varsities (12), Valettas (6), Sunderlands (3), Lincolns (45), Washingtons (12), Shackletons (18), Neptunes (5), Hastings (3), Vampires (36), Meteors (264), Venoms (24), Canberras (48), Sabres (60), Swifts (6), Victor (1), Valiant (1), Vulcan (1), Javelin (1), Hunter (1), Swift (1).

✈ DIMPLES 82

This was the call sign of Boeing B-29 Superfortress, serial number 44-86292, named 'Enola Gay', which was used to drop the first atomic bomb used in anger, on the Japanese city of Hiroshima on 6 August 1945. Enola Gay was the name of the mother of the aircraft's captain, Colonel Paul Tibbets.

✈ GREAT PLANES: LANCASTER

The Avro Lancaster's pugnacious appearance was matched by its performance. It was able to carry a heavy load over long distances and could take a lot of punishment while still being able to bring its crew home. The 'Lanc' took part in the Dam Busters and *Tirpitz* raids, and was able to carry the 22,000lb Grand Slam bomb. Its 33ft-long bomb bay was almost half the total length of the aircraft. The four-engined bomber, which was the mainstay of Bomber Command during the latter part of the war, is regarded as one of the best warplanes of the conflict. Over 7,000 Lancasters were built during the Second World War. A number were converted into civilian passenger transport aircraft called Lancastrians.

This Avro Lancaster is a familiar sight to those attending airshows in the UK. PA474 belongs to the RAF's Battle of Britain Memorial Flight and is the only flying example left in Britain. *Norman H. Ferguson*

 FIRSTS

Event: First untethered flight by humans		
Date: 21 November 1783	**Aircrew:** Jean-François Pilâtre de Rozier; Francois Laurent, the Marquis d'Arlandes	**Craft:** Montgolfier balloon
Notes: The flight was across Paris and lasted around twenty-five minutes.		
First flight in Great Britain		
27 August 1784	James Tytler	Self-built hot-air balloon
Flight was over Edinburgh. Tytler was also editor of the second edition of the *Encyclopaedia Britannica*.		

First flight over English Channel		
7 January 1785	Jean-Pierre Blanchard; John Jeffries	Hydrogen balloon
When Blanchard died after a fall from a balloon, his widow carried flying balloons. She too died in an aerial accident.		
First aviation fatalities		
15 June 1785	Jean-François Pilâtre de Rozier; Pierre Romain	hot-air/hydrogen balloon
Balloon caught fire and crashed while attempting to cross the Channel.		
First powered and sustained flight by heavier-than-air aircraft		
17 December 1903	Orville Wright	Wright Flyer
The flight lasted twelve seconds and covered a distance of 120ft.		
First aeroplane passenger in Europe		
29 May 1908	Aviation sponsor Ernest Archdeacon was piloted by Henry Farman	Voisin-Farman biplane
Flight took place at Issy-les-Moulineaux, near Paris.		
First recognised powered flight in UK		
16 October 1908	Samuel F. Cody	British Army Aeroplane No 1
Cody's flight was made at Farnborough, appropriately enough – later home of the famous airshows.		
First aeroplane crossing of English Channel		
25 July 1909	Louis Blériot	Blériot XI
Blériot won the £1,000 prize offered by the *Daily Mail* for the feat.		
First parachute jump from an aircraft		
1 March 1912	Tony Jannus (pilot); Captain Albert Berry (jumper)	Benoist biplane
When asked if he would repeat his jump, Captain Berry said, 'Never again!'		
World's first scheduled flight with a paying passenger		
1 January 1914	Tony Jannus (pilot)	Benoist Model 14 flying boat
Route was between St Petersburg and Tampa, Florida, and took twenty-three minutes. Abram C. Pheil was the first paying passenger after winning an auction for his ticket.		

First shooting down of another aircraft		
5 October 1914	Sergeant Joseph Frantz (pilot); Corporal Louis Quénault (gunner)	Voisin III

The German Aviatik was downed by Corporal Quénault firing a machine gun from his observer's position in the front of the machine.

First aerial bombing raid on Britain		
19/20 January 1915	German Navy	Zeppelin dirigibles

Great Yarmouth and King's Lynn were attacked, four civilians were killed.

First landing on a moving ship		
2 August 1917	Squadron Commander E.H. Dunning	Sopwith Pup

Dunning landed on HMS *Furious*. Five days later he crashed and was drowned.

First aerial crossing of Atlantic		
27 May 1919	Lt Cdr Albert C. Read (commander/navigator); Lt Walter Hinton (pilot); Lt Elmer F. Stone (pilot) and two flight engineers and a radio operator	Navy-Curtiss NC-4 flying boat

Three aircraft had set off from Newfoundland. Two dropped out of the attempt due to weather. The NC-4s were nicknamed 'Nancies'.

First non-stop aerial crossing of the Atlantic		
15 June 1919	Captain John Alcock (pilot); Lieutenant Arthur Whitten Brown (navigator)	Vickers Vimy

Alcock was killed in an aircraft accident later that year and Brown never flew again.

First circumnavigation of the globe		
28 September 1924	United States Army Air Service pilots	Douglas World Cruiser

Four aircraft took off, only two completed the flight which took five months.+

First flight over North Pole		
12 May 1926	Roald Amundsen, Umberto Nobile (pilot), Oscar Wisting and others	Airship *Norge*
Robert Byrd claimed to have achieved this a few days before in a Fokker Tri-motor but his claims were later found to be unreliable.		
First solo non-stop transatlantic flight		
21 May 1927	Charles Lindbergh	Ryan NYP monoplane *Spirit of St Louis*
Lindbergh became an international celebrity following his 33½-hour flight.		
First Pacific crossing		
9 June 1928	Charles Kingsford Smith, Charles Ulm, James Warner, Harry Lyon	Fokker Tri-motor *Southern Cross*
The flight from Oakland, California to Brisbane, Australia took eighty-three hours' flying time.		
First solo flight around the world		
1933	Wiley Post	Lockheed Vega *Winnie Mae*
Post overcame the disability of only having one eye following an accident. His record-setting flight took seven days, eighteen hours, forty-nine minutes.		
First jet airliner enters service		
2 May 1952	BOAC	de Havilland Comet 1
London to Johannesburg weekly service. Journey took twenty-three hours via Rome, Beirut, Khartoum, Entebbe and Livingstone.		
First man to fly faster than 1,000mph in level flight		
10 March 1956	Peter Twiss	Fairey Delta 2
The average speed of Twiss' two runs was 1,132mph, almost 40 per cent faster than the previous record, held by an American F-100.		
First jet passenger flights across the Atlantic		
4 October 1958	BOAC	de Havilland Comet 4
One Comet flew west to east and another on the opposite route. They passed each other over the Atlantic, 300 miles apart.		

✈ BRITISH TOWN AND VILLAGE SIGNS FEATURING AIRCRAFT

Bassingbourn
Beck Row
Chipping Warden
Coningsby
Debach
East Kirkby
Great Ashfield
Great Bricett
Great Massingham
Hadstock
Haveringland
Hemington
Hunsdon
Lavenham
Litlington
Little Snoring

Little Stukeley
Marham
Matlaske
Methwold
North Thoresby
North Weald Bassett
Oakington and Westwick
Old Warden
Silverstone
Steeple Morden
Stradishall
Thorpe Abbotts
Upavon
Warboys
Weston Colville
Witcombe and Bentham

Hunsdon's village sign shows a Mosquito, which were based at the Hertford-shire airfield. The Amiens prison raid was launched from the station, which closed soon after the end of the war. *Pandaplodder (Wikimedia Commons)*

✈ MESSAGES ON BOMBS

'Extra Havana for Churchill'
Luftwaffe, Second World War

'An Easter Egg for Hitler. Tiger Squadron 2000th sortie'
RAF, Second World War

'Merry Xmas Adolf'
USAAF, Second World War

'Tirpitz it's yours'
Fleet Air Arm, Second World War

'76,389th and last bomb compliments to Charlie from RAAF No. 2 Squadron RAAF Uc Dai Loi'
Royal Australian Air Force, Vietnam, 1971

'Provided for your consumption courtesy of 195(A) Sqn'
RAF, Falklands War

'With love from Weapon Supply HMS Hermes'
Fleet Air Arm, Falklands War

'Who you gonna call? HAS busters!'
RAF, the first Gulf War

✈ USES FOR RETIRED AIRCRAFT

Artificial underwater reef: British Columbia (Boeing 737 airliner)

Artist's studio: Rick Broome, Colorado Springs, Colorado (Boeing 727 cockpit)

Bar: Letka/The Airplane bar, Olomouc, Czech Republic (Tupolev Tu-104) Airways Hotel, Port Moresby, Papua New Guinea (Douglas DC-3) Hi-Flyerz Aviation Bar, Johannesburg (Boeing 747 cockpit)

Bed and Breakfast: Otorohanga, New Zealand (Bristol Freighter)

Boat: 'Cosmic Muffin', Fort Lauderdale, Florida (Boeing 307 Stratoliner)

Cafe: Russia (Antonov An-24)

Children's play area: Sawgrass Mills Mall, Florida (Douglas DC-9)

Cocktail lounge: Penndel, Pennsylvania (Super Constellation)

Disco: Sant Cugat del Vallès, Barcelona (Two Boeing KC-97 Stratotankers)

Fast food outlet: McDonald's, Taupo, New Zealand (Douglas DC-3)

Hostel: Jumbo Stay, Stockholm Arlanda Airport (Boeing 747)

Hotel: Hotel Costa Verde, Costa Rica (Boeing 727)
        Hotelsuites, Teuge, Netherlands, (Ilyushin 18)

House: Ashland City, Tennessee (Douglas DC-8)
        Benoit, Lake Whittington, Mississippi (Boeing 727)
        Laurel, Oregon (Boeing 727)
        Wing House, Malibu, California (Boeing 747 wings)

Hunting lodge: West Leflore Country, Mississippi (Boeing 727)

Mobile kitchen: The Space Shuttle Cafe, California (Douglas DC-3)

Restaurant: DC6 Diner, Coventry Airport (Douglas DC-6)

Restaurant/coffee shop: El Avion, Costa Verde, Costa Rica (Fairchild C-123)

BRAVO NOVEMBER

Bravo November is a Chinook helicopter (registration ZA718) operated by the RAF. Its name stems from the two-letter code it was assigned: BN. It gained recognition in 1982 when it was the only Chinook able to be used in the Falklands conflict. Three others had been sent to the South Atlantic in the *Atlantic Conveyor* cargo ship but were destroyed in an Argentine Exocet missile attack. Bravo November happened to be in the air at the time of the attack.

Squadron Leader Dick Langworthy won a Distinguished Flying Cross medal flying Bravo November during that war. During one night mission supplying the Special Air Service the Chinook flew into the sea. One of the crew was getting ready to rapidly disembark when he found the helicopter out of the water and climbing. Despite some damage, the Chinook was still in flyable condition and continued supplying the British forces until the end of the conflict.

As well as the Falklands the Chinook has seen operational flying in Lebanon, Germany, Northern Ireland, Kurdistan, Iraq and Afghanistan.

✈ THE RAILWAY AIRLINE

On 7 May 1934 the first service of the new Railway Air Services Ltd took place. The airline was set up by the four main railway companies: London Midland & Scottish, London & North Eastern, Great Western Railway and Southern Railway. The airline was intended to link with Imperial Airways services carrying mail and passengers. De Havilland Dragon G-ACPX flew the first flight from Plymouth to Liverpool via Haldon, Cardiff and Castle Bromwich.

✈ INCREDIBLE MACHINES: TARRANT TABOR

26 May 1919 saw the final preparations for the first flight of a new giant of the air. The Tarrant Tabor had three sets of wings with a maximum span of 131ft. The top wing was 37ft above the ground. This six-engined behemoth was the biggest aircraft in the world. The designers and the Royal Aircraft Establishment disagreed over whether it required nose ballast to be added. Eventually a thousand pounds of lead weight was placed in the front of the plane. The Tabor began its take-off run and slowly made its way across Farnborough's grass strip. The two pilots, Captain Frederick Dunn and Captain Percy Rawlings, had only selected full thrust on four of the engines. In order to achieve take-off speed they powered up the two engines on the top wing. This caused the aircraft to immediately nose over, burying the cockpit into the ground. Both pilots died as a result of their injuries. The Tabor did not see operational service.

WORDS OF WISDOM

'The Nazis entered this war under a rather childish delusion that they were going to bomb everybody else and no one was going to bomb them. At Rotterdam, London, Warsaw and half a hundred other places they put that rather naive theory into operation. They sowed the wind, and now they are going to reap the whirlwind.'

Sir Arthur 'Bomber' Harris, chief of RAF Bomber Command, Second World War.

THE FLYING SAINT

St Joseph of Cupertino
Patron saint of air travellers, aviators and astronauts (and of students taking exams).

Joseph, who was also known as 'Joseph the Dunce', earned his nickname after experiencing a series of ecstasies and levitations. It was said that during a procession for Saint Francis of Assisi on 4 October 1630 he rose up and hung above the startled onlookers. During an audience with Pope Urban VIII Joseph levitated into the air and was only able to return to the ground when ordered by one of the Pope's staff. He faced the Inquisition and was subsequently exiled which caused a depression to take hold. Following this his flying stopped. He was warned against appearing in public in case his flights recommenced. He died in 1663 and was canonised in 1767.

Other aviation patron saints include:
Our Lady of Loreto – aircrew
Thérèse of Lisieux – aviators

LUCKY ESCAPES: UNDERWATER EJECTION

In 1969, US Navy pilot Russ Pearson was nearing the end of his qualifying period as a pilot able to land on aircraft carriers. During a night landing his A-7 Corsair jet fell over the side of the aircraft carrier's deck. Pearson only had a few seconds before the water pressure would make escape impossible. He had only one option: pulling the ejector seat handle. The rocket motor wasn't able to fire him clear of the water but was enough to break the canopy and get him free of the sinking aircraft. Pearson was still in trouble as he was disorientated in the darkness and his parachute was dragging him under. Eventually he was able to free himself and inflate his life vest. The deck marshals helped by throwing their waterproof wands into the water to mark his position. Pearson was picked up by a rescue helicopter and returned to the carrier. His luck held when a medical evacuation flight that he had been scheduled to take, but did not travel on, crashed into the sea, killing all on board. He retired in 1992.

✈ 21 NOVEMBER 1783

On this day the first untethered flight by humans took off from Paris' Chateau de la Muette. The men in the Montgolfier balloon were Jean-François Pilâtre de Rozier and Francois Laurent, the Marquis d'Arlandes. The Marquis described this historic flight:

'We set off at fifty-four minutes past one o'clock. The balloon was so placed that M de Rozier was on the west and I was on the east. The machine, says the public, rose with majesty. I think few of them saw that, at the moment when it passed the hedge, it made a half-turn and we changed our positions, which, thus altered, we retained to the end. I was astonished at the smallness of the noise and motion among the spectators occasioned by our departure. I thought they might be astonished and frightened and might stand in need of encouragement so I waved my arm, with small success. I then drew out and shook my handkerchief, and immediately perceived a great movement in the yard. It seemed as if the spectators all formed one mass, rushing by an involuntary motion towards the wall, which they seemed to consider as the only obstacle between us.

'At this time M de Rozier said, "You are doing nothing and we are not mounting." "Pardon me," I replied. I threw a truss of straw upon the fire, stirring it a little at the same time, and then quickly turned my face back again, but I could no longer see La Muette. Astonished I looked at the river. I perceived the confluence of the Oise. And naming the principal bends of the river by the places nearest them, I cried, "Passy, St. Germain, St. Denis, Sèvres!"'

The balloon continued its journey across Paris for some 5½ miles. At one point an on-board fire had to be distinguished. As the flight neared its end, they discovered their flight path was taking them towards some mills:

'I threw a bundle of straw on the fire, and shaking it in order to make it inflame more easily, we rose, and a new current carried us a little towards

The highly decorated Montgolfier hot-air balloon depicted as it lifted off on its historic flight. The balloon displayed signs of the zodiac, the King's face and eagles' wings. *Library of Congress.*

our left. M de Rozier said again "Take care of the mills" but as I was looking through the aperture of the machine, I could observe more accurately that we could not meet with them and said, "We are there". The moment we touched the ground, I raised myself up in the gallery, and perceived the upper part of the machine to press very gently on my head. I pushed it back and jumped out of the gallery, expected to find it distended, but was surprised to find it perfectly emptied and quite flattened.'

✈ CURRENT JOBS IN THE RAF

Aircrew: Pilot, Weapon Systems Operator, Weapon Systems Operator (Linguist). Technical and engineering: Aircraft Technician (Avionics), Aircraft Technician (Mechanical), Engineer Officer, General Technician Electrical, General Technician Mechanical, General Technician Workshops, Survival Equipment Fitter, Weapon Technician.
Force protection: Firefighter, RAF Police, RAF Police Officer, RAF Regiment Gunner, RAF Regiment Officer.
Medical and medical support: Biomedical Scientist, Dental Nurse, Dental Officer, Dental Technician, Environmental Health Technician, Medical Officer, Medical Support Officer, Medical Support Officer (Physiotherapist), Nursing Officer, Operating Department Practitioner, Pharmacy Technician, Radiographer, RAF Medic, Registered Nurse (Adult), Registered Nurse (Mental Health), Student Nurse (Adult).
Personnel support: Chaplain, Legal Officer, Musician, Personnel Officer, Personnel Support, Physical Training Instructor.
Air operations support: Aerospace Battle Manager, Aerospace Systems Operator, Air Cartographer, Air Traffic Control Officer, Air Traffic Controller, Flight Operations Assistant, Flight Operations Officer.
Communications and intelligence: ICT Aerial Erector, ICT Technician, Intelligence Analyst, Intelligence Analyst (Voice), Intelligence Officer, Photographer.
Logistics: Caterer, Chef, Driver, Logistics Officer, Mover, Supplier.

✈ YOU CAN CALL ME BETTY

During the Second World War Japanese warplanes were given English codenames as it was felt the Japanese classification system was too complicated, and could confuse Allied service personnel. Fighters were given male names and all other types, except for single-engined reconnaissance aircraft, were given female names. Some of the aircraft had Japanese names which are listed along with their English translations:

Babs — Mitsubishi Ki-15
Betty — Mitsubishi G4M
Frank — Nakajima Ki-84 Hayate (Gale)
Emily — Kawanishi H8K
George — Kawanishi N1K-J Shiden (Violet Lightning)
Grace — Aichi B7A Ryusei (Shooting Star)
Jack — Mitsubishi J2M Raiden (Thunderbolt)
Jake — Aichi E13A
Judy — Yokosuka D4Y Suisei (Comet)
Kate — Nakajima B5N
Lorna — Kyushu Q1W Tokai (Eastern Sea)
Mavis — Kawanishi H6K
Nell — Mitsubishi G3M Rikko
Nick — Kawasaki Ki-45 Toryu (Dragon Slayer)
Norm — Kawanishi E15K Shiun (Violet Cloud)
Oscar — Nakajima Ki-43 Hayabusa (Peregrine Falcon)
Paul — Aichi E16A Zuiun (Auspicious Cloud)
Peggy — Mitsubishi Ki-67 Hiryu (Flying Dragon)
Randy — Kawasaki Ki-102
Sally — Mitsubishi Ki-21
Tony — Kawasaki Ki-61 Hien (Swallow)
Val — Aichi D3A
Zeke — Mitsubishi A6M (commonly called 'Zero' due to its Japanese designation as 'Mitsubishi Navy Type 0 Carrier Fighter')

✈ DID YOU KNOW?

14,166:
The number of British pilots killed in the First World War. Over 8,000 died in training, more than were lost in actual combat.

✈ INCREDIBLE MACHINES: ILYA MUROMETS

This impressive series of machines was named after a tenth-century Russian mythological hero. They were designed by Igor Sikorsky who had built the world's first four-engined aircraft which he developed into the Ilya Muromets. The aircraft was a four-engined biplane, 57ft long with a wingspan of 113ft. This made it, at the time of its introduction in 1913, the biggest aircraft in the world but it wasn't just its size that made it remarkable. It was designed to be a luxurious passenger aircraft and came equipped with a dining room, toilet, central heating, electric lighting and a viewing deck where passengers could stand outside and take in the view as they flew along. Its maximum speed was below 70mph.

At the outbreak of the First World War Tsar Nicholas II ordered them to be built for military service and they carried out the first ever heavy bombing missions. It was said that German fighter pilots were nervous about attacking due to the plane's strong defences – the Ilya Muromets had the first tail gunner position. Sikorsky later immigrated to America and set up a company building helicopters.

PLANES WITH FORWARD-SWEPT WINGS

Grumman X-29: Experimental jet

Sukhoi Su-47 Berkut: Fighter

Hamburger Flugzeugbau Hansa Jet: Civilian business jet

Junkers 287: Second World War experimental jet

The X-29 was built as a technology demonstrator and was found to have good aerodynamic properties, despite its unusual appearance. *NASA*

Cornelius Mallard: Civilian light aircraft

Cornelius XFG-1: Fuel tanker glider

OKB-1 EF 140: Russian experimental bomber

 MYSTERIES OF THE SKY: FLIGHT 19

The Bermuda Triangle is an area drawn between points in Florida, Bermuda and Puerto Rico. Various unexplained events in it have led to much speculation. Ships and aircraft have disappeared while travelling through the Triangle and one of the most talked-about instances is Flight 19. Five American TBM Avenger torpedo bombers were on a training flight in 1945 when they vanished without trace. The aircraft had taken off from Fort Lauderdale in Florida on a pre-arranged training flight. Radio transmissions were heard that suggested the planes became lost. Another aircraft was launched to look for the flight, but it too disappeared. No trace of the Avengers or their crews has ever been found. Were they abducted by aliens? Did they fly into a worm-hole vortex taking them to another dimension? Or did the flight leader misread their position and lead them the wrong way until they ran out of fuel and crashed into the sea? Whatever the reason, all fourteen men of Flight 19 and the thirteen sent to find them died in one of aviation's most enduring mysteries.

SPITFIRE VS HURRICANE

	Spitfire IA*	Hurricane I*
Wingspan	36ft 10in	40ft
Length	29ft 11in	31ft 5in
Armament	8 x .303 Browning machine guns	8 x .303 Browning machine guns
Engine	Rolls-Royce Merlin II/III	Rolls-Royce Merlin II/III
Maximum speed (mph)	355	316
Service ceiling (feet)	34,000	33,200
Time to altitude (15,000ft) in minutes	6.2	6.3

Numbers in squadron service (during Battle of Britain)	957	1,326
Percentage of aircraft shot down in Battle of Britain by both types	45 per cent	55 per cent
Total numbers built (all marks)	22,759 (including Seafires)	14,533

* Battle of Britain variants

The Spitfire's cleaner aerodynamic shape and famous elliptical wings are shown against those of the Hurricane in this photograph of two of the aircraft being flown in recent times. *SAC Neil Chapman – Crown copyright (Open Government Licence)*

✈ THOSE WHO SERVED

Hollywood actor James Stewart is perhaps the most famous personality to have served in an air force during the Second World War or during National Service. Despite the reputation of being 'luvvies' many actors answered the call of duty, including perhaps the luvviest of all, Laurence Olivier:

(Second World War unless stated)

UNITED KINGDOM

Royal Air Force

Aircrew:

Ken Adam (production designer)	pilot
Richard Attenborough (actor/director)	air gunner/cameraman
Richard Burton (actor)	navigator (National Service)
Roald Dahl (writer)	fighter pilot
Denholm Elliot (actor)	radio operator/gunner
Warren Mitchell (actor)	navigator (National Service)
Patrick Moore (astronomer)	navigator
Donald Pleasence (actor)	wireless operator/gunner
Norman Tebbit (politician)	pilot (National Service)
Kenneth Wolstenholme (sports commentator)	bomber pilot

Other trades:

Pam Ayres (poet)	Women's Royal Air Force
David Bailey (photographer)	National Service
Alan Bates (actor)	National Service
Alec Bedser (cricketer)	RAF Police
Michael Bentine (comedian)	intelligence officer
Max Bygraves (entertainer)	fitter
Arthur C. Clarke (science fiction writer)	radar specialist
Ronnie Corbett (comedian)	National Service
Bruce Forsyth (entertainer)	National Service
Tony Hancock (comedian)	RAF Regiment
Robert Hardy (actor)	National Service
Rex Harrison (actor)	flight controller
Roy Hudd (actor/comedian)	National Service

Christopher Lee (actor)	intelligence officer
Ken Loach (film director)	National Service
Dan Maskell (tennis commentator)	rehabilitation officer
Bob Monkhouse (comedian)	National Service
Des O'Connor (entertainer)	National Service
Peter Sellers (actor)	concert entertainer
Eric Sykes (comedian/writer)	wireless operator
Fred Trueman (cricketer)	National Service
Bill Wyman (rock musician)	National Service

Royal Navy Air Service/Fleet Air Arm/Royal Naval Volunteer Reserve

George Martin (music producer)	pilot
Laurence Olivier (actor)	pilot
Ralph Richardson (actor)	pilot

UNITED STATES OF AMERICA

United States Army Air Forces (unless stated)

Gene Autry (musician/actor)	cargo pilot
James Best (actor, *Dukes of Hazzard*)	B-17 radio operator
Richard Boone (actor)	air gunner (US Navy)
Charles Bronson (actor)	air gunner
Johnny Cash (musician)	radio operator (1950s)
William Conrad (actor, *Cannon*)	fighter pilot
Clark Gable (actor)	air gunner
Charlton Heston (actor)	radio operator/gunner
George Roy Hill (film director)	pilot (Korean War)
Brian Keith (actor)	air gunner (US Marine Corps)
Karl Malden (actor)	non-commissioned officer
Walter Matthau (actor)	radio operator/gunner
Burgess Meredith (actor)	officer
Paul Newman (actor)	radioman/gunner (US Navy)
Jack Palance (actor)	pilot
Tyrone Power (actor)	pilot (US Marine Corps)
Gene Roddenberry (*Star Trek* creator)	pilot
Robert Stack (actor)	gunnery instructor (US Navy)
James Stewart (actor)	bomber pilot
Robert Taylor (actor)	flying instructor (US Navy)
Dennis Weaver (actor)	fighter pilot (US Navy)

CANADA

Royal Canadian Artillery

James Doohan (actor, *Star Trek)*   pilot

JAPAN

Imperial Japanese Air Force

Toshiro Mifune (actor)   aerial photographer

Brigadier General James M. Stewart here pictured in his US Air Force uniform. As well as Second World War, the Hollywood star also flew on missions in the Vietnam War. *USAF*

AN IRISH AIRMAN FORESEES HIS DEATH

'I know that I shall meet my fate
Somewhere among the clouds above;
Those that I fight I do not hate,
Those that I guard I do not love;
My country is Kiltartan Cross,
My countrymen Kiltartan's poor,
No likely end could bring them loss
Or leave them happier than before.
Nor law, nor duty bade me fight,
Nor public men, nor cheering crowds,
A lonely impulse of delight
Drove to this tumult in the clouds;
I balanced all, brought all to mind,
The years to come seemed waste of breath,
A waste of breath the years behind
In balance with this life, this death.'

W.B. Yeats, 1919.

DID YOU PACK YOUR BAG YOURSELF?

Compass, scientific instruments, brandy, silk ensigns, biscuits, cork jackets, letters to French nobility.

– Items carried on first-ever balloon flight over the English Channel, 7 January 1785.

 MINE'S A PINT

A few aviation-themed beers:

Airdale Afterburner
Aviator Red
B-25 Brown Ale
Battle of Britain Ale
Captain Alf
Dakota Ale
Halifax Bomber

Hurricane Ale
Lancaster Bomber
Meteor
Mitchell's Dream
Mosquito Photo Recce
P-47 Warbird Wheat
P-51 Pale Ale

Red Hunter
Spitfire Ale

Squadron Scramble
Wooden Wonder

✈ THE AERIAL SHOWMAN

The first powered flights in Britain were made by Samuel F. Cody. Cody was a colourful character, an American who had made his name in Wild West shows before turning his attention to flying machines. He was based at the army station at Farnborough, where he worked on building an aircraft known as the British Army Aeroplane Number 1. On 16 October 1908 Cody ensured his place in the history books with the first powered and sustained flight in Britain in a heavier-than-air machine. *The Times* reported on the historic event:

Samuel F. Cody was one of the most colourful figures in the early days of aviation. He had been born Cowdery but changed his name and based his appearance on that of William F. 'Buffalo Bill' Cody, who also ran Wild West shows. *Library of Congress*

'Starting from the brow of the hill, he lifted the machine, after a run of a few paces and in less than 100 yards he had reached an altitude of between 30 and 40 feet, so that the spectators on the southern boundary of the common saw the aeroplane flying above the roof of the balloon shed. Shaping a course that led to Cove Common, Mr Cody kept the aeroplane about 30 feet from the ground and he travelled across the greensward for a quarter of a mile. Then he was confronted by a group of birch trees 30 feet high. The aeroplane cleared these with ease. Beyond them, however, lay a stretch of country dotted with gorse and woodland. Mr Cody evidently had no mind to face these obstacles, for the aeroplane was suddenly wheeled round on its left wing at an acute angle. This evolution was fatal to further progress, for the left wing dipped, the aeroplane itself lowered and the small cycle wheel on the left extremity of the cross spar struck the ground with such force that the timber snapped.'

Cody might have crashed, but he had proven the effectiveness of his machine. Unfortunately the authorities were not so impressed and cancelled his work a year later, seeing the future of military aviation in airships. Cody said after his historic flight, 'I am sorry that the accident occurred, but I have accomplished what I aimed at. I have constructed a machine which can fly.'

✈ BOMB NAMES

Cookie	–	'blockbuster' type of 4,000, 8,000 and 12,000 pounds
Daisy Cutter	–	15,000-pound bomb, used in Vietnam for ground clearing
Fat Man	–	atomic bomb, used on Nagasaki
Grand Slam	–	22,000-pound 'Earthquake' bomb
Highball	–	'bouncing bomb' proposed for use against shipping
Little Boy	–	atomic bomb, used on Hiroshima
Tallboy	–	10-ton 'Earthquake' bomb
Upkeep	–	'bouncing bomb' used in Dams Raid

✈ PHANTOM FLYERS

Many ghostly happenings have been reported in the world's skies. Here are a few of the spookiest:

Montrose
Montrose became the first operational airfield in Britain when the Royal Flying Corps' 2 Squadron arrived in 1913 and it is the fate of one of that

squadron's pilots that gives the Angus airfield its place in the paranormal world. Lieutenant Desmond Arthur, from County Limerick in Ireland, was flying over the airfield in a BE.2 biplane when it suffered catastrophic damage to the wing. The aircraft broke up and Arthur fell free of the machine to his death. A few years later, in the middle of the First World War, several officers started seeing a pilot standing near the fire in their bedrooms, or approaching the officers' mess, but who strangely disappeared. The sighting became known as the Irish Apparition, as a visitor recognised the ghost as Arthur. In 1963 one of the most bizarre of experiences was reported by Sir Peter Masefield, a respected figure who had helped develop British aviation. He wrote of how he was about to fly south from the Moray Firth in his Chipmunk when an Irish-speaking pilot, dressed in old flying clothing, asked for a lift. Masefield took off and headed south, his passenger remaining silent throughout the flight. Their route took them near Montrose and it was there that a biplane appeared. Masefield watched horrified as the plane broke up in the air and plummeted towards the ground, the pilot falling free, his arms and legs flapping despairingly. Masefield heard a wail from the back seat. He looked round to see the seat empty, his passenger gone. He landed immediately at Montrose's airfield. No one had seen this other aircraft. Masefield's account was published before Christmas and whether it was a true account of a real event or Yuletide ghost story remains unknown ...

Flight 401

In 1972 an Eastern Airlines Lockheed L-1011 TriStar wide-body airliner crashed into the Florida Everglades. The accident killed 101 passengers and crew. The story would probably have ended there were it not for some unusual events in the months afterwards. Crew members and passengers on other Eastern flights reported seeing and speaking to two flight officers in Eastern uniforms. They were identified as Captain Bob Loft and flight engineer Don Repo ... both of whom had died in the crash. There were numerous further sightings of the pair. Some witnesses reported the figures vanishing in front of their eyes. The spectres were mainly spotted on aircraft that had reused salvaged parts from 401's aircraft. Eastern denounced the sightings, but removed the spare parts, and as mysteriously as they had begun, the sightings stopped.

Deke's Plane

Deke Slayton was one of the original Mercury 7 astronauts, chosen to be the first Americans into space. Unfortunately medical issues prevented him getting into orbit until 1975. He retired from NASA after a notable career, and died in 1993. It was a sad passing but otherwise unremarkable until it was reported his personal racing aeroplane, serial number N21X, had violated noise abatement regulations after taking off from John Wayne Airport in California on the morning of his death. Slayton's widow was surprised at being told of this as her husband had died five and a half hours before the alleged incident, and the aircraft was in a museum a few hundred miles away. The distinctive red aircraft was reported as being seen heading off towards the Pacific Ocean.

WHAT WERE THEY THINKING?

Some of the more bizarrely named aircraft:

BAT Baboon
Blackburn Blackburd
Boeing P-26 Peashooter
Bristol Tramp
Christmas Bullet

De Havilland Spider Crab
De Pischoff 1907
Short Shirl
Sopwith Snail
Supermarine Baby

SLOW (FLYING) BOAT TO AUSTRALIA

In the 1930s flying boats were preferred for certain long-range routes as long runways or airfields were not always suitable or available in many of the countries visited. With this in mind British aircraft manufacturer Short built the S.23/30/33 series, known as the Empire flying boats. They were designed to carry mail and passengers to the far reaches of the Empire. Most were operated by Imperial Airways on routes to Africa, Asia and Australia, while others were flown by Australian and New Zealand operators.

The Imperial Airways route to Australia would appear torturously protracted to any modern traveller. The following route from 1938 had ten overnight stops. The trip started with trains taking passengers from London to Southampton to board the Short Empire aircraft before they headed off on the following journey:

Southampton – Marseilles (France) – Rome (Italy) – Brindisi (Italy) – Athens (Greece) – Alexandria (Egypt) – Tiberias (Palestine) – Habbaniyeh (Iraq) – Basra (Iraq) – Bahrain – Dubai (Oman) – Karachi (India) – Rajsamand (India) – Gwalior (India) – Allahabad (India) – Calcutta (India) – Akyab (Burma) – Rangoon (Burma) – Bangkok (Siam) – Penang (Malaya) – Singapore [from here the route was flown by Qantas, using same type of aircraft] – Batavia (Netherlands Indies) – Sourabaya (Netherlands Indies) – Koepang (Netherlands Indies) – Darwin (North Australia) – Karumba (Queensland) – Townsville (Queensland) – Gladstone (Queensland) – Brisbane (Queensland) – Sydney (New South Wales)

DID YOU KNOW?

22:
The average age of aircrew in RAF Bomber Command during the Second World War.

AIRPORTS NAMED AFTER PEOPLE

Bob Hope Airport, California, USA
Charles de Gaulle, Paris, France
Chinggis Khaan International Airport, Ulaanbaatar, Mongolia
George Best Belfast City Airport, UK
Giuseppe Verdi Airport, Parma, Italy
John F. Kennedy International Airport, New York, USA
John Paul II International Airport Kraków-Balice, Poland
John Wayne Airport, California, USA
Liverpool John Lennon Airport, UK
Skopje Alexander the Great Airport, Macedonia

HARRY FERGUSON

Ferguson's name is synonymous with tractors, the Irishman designing an implement coupling device that revolutionised mechanised farming. His famous 'Little Grey Fergie' tractors were manufactured in their hundreds of thousands in the mid-twentieth century and can still be seen in operation today. What is perhaps less known is Ferguson's aviation claim to fame: he was the first man to fly in Ireland. On 31 December 1909 the man from County Down took to the air in a monoplane at Belfast's Hillsborough Park. Ferguson had travelled to air displays in England and France to familiarise himself with this new form of transport before building his own. Not content with an engineering career involving agriculture and aviation, Ferguson later went on to develop a four-wheel-drive chassis for racing cars. Driven by Stirling Moss, the Project 99 won a Formula One race in 1961.

WORDS OF WISDOM

'There are old pilots. There are bold pilots. There are no old, bold pilots.'

Anon.

 FLIGHT ON FILM

The Hijack/Heist/Panic at 30,000ft Movie:
Air Force One
Con Air
Crash Landing
Flightplan
Panic in the Skies

Passenger 57
Skyjacked
Snakes on a Plane
Turbulence
Turbulent Skies

The Intrepid Aviator Movie:
Fly Away Home
The Great Waldo Pepper
The Right Stuff

The Sound Barrier
The Spirit of St Louis
Test Pilot

The Airport Disaster Movie:
Airport
Airport 1975

Airport '77
The Concorde – Airport '79

The Shot Down or Crashed and Now in Peril Movie:
Alive
Bat 21
Behind Enemy Lines
Black Hawk Down

Flight of the Intruder
Flight of the Phoenix
One of Our Aircraft is Missing
Rescue Dawn

The Dogfighting over the Trenches Movie:
Aces High
The Blue Max
Dawn Patrol
Flyboys

Hell's Angels
The Red Baron
Von Richthofen and Brown
Wings

The Flight Number Movie:
Crash: The Mystery of Flight 1501
Falling from the Sky: Flight 174
Flight 90: Disaster on the Potomac
Flight 93
Hijacked: Flight 285
Mercy Mission: The Rescue of Flight 771

Murder on Flight 502
Rough Air: Danger on Flight 534
The Disappearance of Flight 412
The Ghost of Flight 401
The Tragedy of Flight 103: The Inside Story
United 93

The Comedy Movie:
Airplane!
Hot Shots!

Those Magnificent Men in Their Flying Machines, or How I Flew from London to Paris in 25 hours 11 Minutes

The Traditional War Movie:

633 Squadron	Memphis Belle
Angels One Five	Mosquito Squadron
Battle of Britain	Reach for the Sky
The Dam Busters	Tora! Tora! Tora!

The Gung-Ho American Movie:

A Gathering of Eagles	Strategic Air Command
Flying Leathernecks	Thirty Seconds Over Tokyo
Flying Tigers	Top Gun
Pearl Harbor	Twelve O'Clock High

The Aircraft Carrier Goes Back in Time and Could Change the Course of History at Pearl Harbor Movie:
The Final Countdown

✈ AIR FORCE MOTTOS

Air Force	Latin (or native language)	English
Finnish Air Force	Qualitas potentia nostra	Quality is our power
French Air Force	Faire face	Face your fears
Indian Air Force	Nabha sprisham deeptam	Touching the sky with glory
Italian Air Force	Virtute siderum tenus	With valour to the stars
Royal Australian Air Force	Per ardua ad astra	Through struggle to the stars
Royal Canadian Air Force	Sic itur ad astra	Such is the pathway to the stars
Royal Norwegian Air Force	Konge folk og fedreland	For King, People and Fatherland
Pakistan Air Force	Sahrast ke daryast tah-e-bal-o-par-e-mast	Lord of all I survey
Portuguese Air Force	Ex mero motu	Of his own free will
Royal Air Force	Per ardua ad astra	Through adversity to the stars
United States	n/a	Aim high ... Fly–Fight–Win

MYSTERIES OF THE SKY: GLENN MILLER

Glenn Miller was one of America's foremost band leaders. When his country entered the Second World War he joined up and set up the Glenn Miller Army Air Force Band which performed morale-raising concerts for the troops. It was on his way to arrange one such appearance that Miller disappeared. On 15 December 1944 his Norseman plane took off from RAF Twinwood Farm, bound for Paris. The plane never arrived. What had happened? Like all unexplained events of this nature theories with varying attachments to plausibility have been aired. They include that his plane was downed through engine failure, it suffered ice on the wings, it was hit by Allied anti-aircraft fire, or it was destroyed by jettisoned bombs from a RAF Lancaster flying at a higher altitude. Another more outlandish claim was that Miller made it to Paris where he was later killed in a brothel.

DID YOU KNOW?

2,029,399:
Total number of aviation photographs displayed on website Airliners.net (as at 31 July 2012).

DEFUNCT BRITISH AIRCRAFT MANUFACTURERS

ABC
Airco
Airspeed
Alliance
ANEC
Armstrong Whitworth
Arpin
Arrow
ASL
Auster
Austin
AV Roe
Aviation Traders
BAMC
BAT
Beagle

Beardmore
Blackburn
Boulton Paul
Bristol
Britten-Norman
Carden-Baynes
Central
Chilton
Chrislea
Cierva Autogiro
Comper
Cunliffe-Owen
Dart
De Bolotoff
de Havilland
Deekay

Edgley
English Electric
Fairey
Folland
Foster Wikner
Gloster
Grahame-White
Handley Page
Hawker
Hendy
Heston
Hillson
Hunting
London & Provincial
Luton
Marendaz
Martin Baker
Martinsyde
Miles

NDN
Parnall
Percival
Poboy
Portsmouth
Royal Aircraft Factory
Saunders Roe
Scottish Aviation
Short
Slingsby
Sopwith
Southern
Spartan
Supermarine
Trago Mills
Vickers Armstrong
Westland
White & Thompson

✈ PEOPLE AND PLACE PLANE NAMES

Heavy bombers in the Second World War were named after British cities, leading to the Avro Lancaster, Avro Lincoln, Avro Manchester, Handley Page Halifax and Short Stirling. Flying boats were given names of coastal towns or ports: Saunders-Roe Lerwick, Saunders-Roe London, Short Shetland, Short Seaford, Short Sunderland, Supermarine Scapa, Supermarine Southampton and Supermarine Stranraer. Maritime aircraft were named after naval figures or explorers, so there was the Avro Anson (Admiral George Anson, who circumnavigated the globe), Lockheed Hudson (Henry Hudson, who discovered the Hudson river), Avro Shackleton (Ernest Shackleton, the Antarctic explorer) and the Bristol Beaufort (Francis Beaufort, inventor of the wind measurement scale). Training aircraft were assigned names stemming from academic institutions or names for teachers: Airspeed Oxford, Avro Prefect, Avro Tutor, North American Harvard, Boulton Paul Balliol, de Havilland Dominie, de Havilland Don, Miles Magister, Miles Master, Miles Mentor, Percival Proctor, Percival Prentice and Percival Provost. Fighter aircraft built by Hawker received names based on wind, hence the Hurricane, Typhoon and Tempest. The Spitfire's name came from the manufacturer's chairman who was reported to have used the term to describe his daughter!

VULTURE SQUADRON

In the children's TV show *Dastardly and Muttley in Their Flying Machines*, Vulture Squadron, comprising Dick Dastardly, Muttley, Zilly and Klunk, were tasked with stopping the communication bird Yankee Doodle Pigeon. They used a variety of methods in their attempts:

Birdseed/locomotive tracks	failure
Boulder drop	failure
Bucket/heavy weight	failure
Butterfly net/piano	failure
Butterfly nets	failure
Ding Dong (large gorilla)	failure
Diving board/parachute	failure
Egg beater contraption	failure
Ejector seats	failure
Flypaper	failure
Magnet/anvil	failure
Mechanical flying pigeon	failure
Mousetrap	failure
Oversized flat irons	failure
Oversized tennis rackets	failure
Parachute retardation	failure
Pepper shell/Sneezo-Sniffer Missile	failure
Pigeon grabber	failure
Pigeon hammer	failure
Pigeon pulveriser	failure
Piledriver	failure
Spring-powered aeroplane	failure
Steel jaws	failure
Tail chase	failure
... and finally ...	
Diving board/aerial pool	initial success; ultimate failure

FAMOUS FEATS OF AVIATION: ENGLISH CHANNEL CROSSING

The *Daily Mail* had set a prize of £1,000 for a flight 'across the channel between England and France to be accomplished in daylight without touching the sea'. In July 1909 two French aviators were poised to make their attempts. Louis Blériot had burnt his foot on an exhaust pipe and was on crutches but was still able to operate the rudder, and was determined to win. His rival, Hubert Latham, had ditched in the sea in a previous attempt but was racing to repair his machine. On the morning of 25 July 1909, Blériot rose early to make his bid for glory. He later described his experiences:

'At 4:30 daylight had come … A light breeze from the southwest was beginning to blow. The air was clear. Everything was prepared. I was dressed in a khaki jacket lined with wool for warmth over tweed clothes and beneath my engineer's suit of blue cotton overalls. My close fitting cap was fastened over my head and my ears.

'I had neither eaten nor drunk anything. My thoughts were only upon the flight and my determination to accomplish it this morning. At 4:35 the signal is given, and in an instant I am in the air, my engine making 1,200 revolutions, almost its highest speed, in order that I may get quickly over the telegraph wires along the edge of the cliff. As soon as I am over the cliff I reduce my speed. There is now no need to force my engine. I begin my flight steady and sure toward the coast of England. I have no apprehensions, no sensations, *pas du tout*.

'I am alone. I can see nothing at all. For ten minutes I am lost. It is a strange position to be alone, unguided, without a compass in the air over the middle of the channel. I touch nothing. My hands and feet rest lightly on the levers. I let the aeroplane take its own course. I care not whither it goes. For ten minutes I continue, neither rising nor falling nor turning, and then twenty minutes after I have left the French coast I see the green hills of Dover, the castle, and away to the west the spot where I intended to land.

'What can I do? It is evident that the wind has taken me out of my course. I am almost west of Margaret's Bay, and I am going in the direction of the Goodwin Sands. Now it is time to attend to steering. I press a lever with my foot and turn easily toward the west, reversing the direction in which I am now traveling. Now, indeed, I am in difficulties, for the wind here by the cliffs is much stronger and my speed is reduced as I fight against it, yet my beautiful aeroplane responds.

'Once more I turn my aeroplane, and describing a half-circle I enter the opening and find myself again over dry land. Avoiding the red buildings on my right, I attempt a landing, but the wind catches me and whirls me around two or three times. At once I stop my motor, and instantly my machine falls upon the land from a height of 65ft. In two or three seconds I am safe upon your shores. Soldiers in khaki run up, and a policeman and two of my

This photograph was taken from a boat following the progress of Louis Blériot as he flew across the English Channel. The image captures him as he nears the cliffs of Dover. *George Grantham Bain Collection/Library of Congress*

compatriots are on the spot. They kiss my cheek. The conclusion of my flight overwhelms me. I have nothing to say, but accept the congratulations.'

Latham had missed his chance, remaining asleep while Blériot was on his way. He attempted to overshadow Blériot's flight by crossing the Channel and continuing on to London, but this attempt failed when he ditched a few miles short of Dover.

✈ LUCKY ESCAPES: CHUCK YEAGER

Charles 'Chuck' Yeager was the first recorded pilot to successfully navigate the sound barrier when he flew the rocket-powered Bell X-1 to Mach 1.06 in 1947. He remained a test pilot at Edwards Air Force Base where one of the types he flew was the NF-104A, a modified Lockheed Starfighter fitted with an additional rocket motor producing 6,000 pounds of thrust. With this extra power the NF-104A could reach altitudes above 100,000ft.

On 10 December 1963 Yeager took off as normal, fired the rocket and zoom-climbed to 104,000ft when the aircraft lost control. It fell in a flat spin. Yeager later said it performed fourteen complete rotations, and he ejected on the thirteenth. He fell free of his rocket-powered seat, but the parachute didn't open immediately. The seat, which was now falling at the same speed and still had its rocket propellant burning, became tangled in the parachute lines and smashed into Yeager's crash helmet. His pressurised suit was continuing to pump oxygen into the helmet and when it ignited, acted like a blow torch. Yeager managed to open the helmet's faceplate and the supply was turned off. He survived, but had suffered severe burns to his hands and face. He later went on to lead a combat squadron in the Vietnam War and retired from the US Air Force in 1975.

✗ FAMOUS FEATS OF AVIATION: ALCOCK AND BROWN

In 1919 an Englishman and a Scotsman went into an Irish bog. Not the start of a ropey joke but the end of the journey of Captain John W. Alcock (pilot, Englishman) and Lieutenant Arthur Whitten Brown (navigator, Scotsman). They had achieved the first non-stop crossing of the Atlantic, flying in a twin-winged Vickers Vimy. They had left Newfoundland in the afternoon of 14 June and sixteen hours later landed outside Clifden in Galway. En route they encountered thick clouds, snow and hail and navigation proved to be difficult. The Vimy dropped perilously close to the sea following a spin from 4,000ft. Unlike the American crossing the previous month, the British crew had no line of support ships available. It is said Brown never flew again after this trip and it's hard not to see why. Louis Blériot later wrote: 'In the Great Book of the conquest of the air their exploit must be inscribed among the most successful and most daring, and also as one of the most sagely carried out.'

DID YOU KNOW?

7,000:
The number of aircraft estimated by the Federal Aviation Authority to be in the air over America at any given time.

YOUR COUNTRY NEEDS YOU

Some slogans from RAF recruitment posters:

'Royal Air Force. The Gateway to a Brighter Life.'
'The Best Life of All!'
'Be an Airman.'
'Let's Go – Wings for Victory.'
'The call of the east – the RAF answers it.'
'Get a good rise by joining the RAF. Pay 3/ to 18/ a day.'
'Join an aircrew in the RAF.'
'Join the Royal Air Force and make a direct hit.'
'Handsome men join today. The sky awaits you. Aircrew.'
'Highway for youth as an Apprentice in the Royal Air Force.'
'For a really adventurous life – be trained as an Air Observer. Openings are now available for sound, responsible young men. Join the RAF.'
'Fly it.'
'Supply it.'
'Join the Royal Air Force and share their Honour and Glory. Age 18 to 50. Rates of pay from 1/6 to 12/- per day. If you join the Royal Air Force voluntarily, you cannot be transferred to the Army or Navy without your own consent.'
'Hard hitters are these men – pilots of the RAF. The bigger the job, the keener they are to tackle it. Both in Europe and Africa they are hammering the enemy. You can share in their splendid work. You too can hit back for Britain. Volunteer for flying duties – do it now! Fly with the RAF.'

Woman's Royal Air Force

'Women! The Royal Air Force needs your help! Serve your country by joining the WRAF. There is fit work for every fit woman. Woman's Royal Air Force.'

'This girl's going places in the Woman's Royal Air Force.'

'British Women! – the Royal Air Force needs your help as clerks, waitresses, cooks, experienced motor cyclists and in many other capacities, full particulars from the nearest employment exchange. Enrol at once in the WRAF. Woman's Royal Air Force.'

'There's a place for you ... in the WRAF.'

 ROCK STAR AIRLINES

While rock bands start off their careers having to share vans, if they make it big they can have access to their own aircraft, either owned or hired for the duration of their tours. These are some of the rock gods' sky chariots:

Band/Singer	Aircraft	Registration
Aerosmith	Gulfstream II	N267PS
The Beatles	Boeing 707	N704PA 'Jet Clipper Beatles'
Coldplay	Embraer EMB-135ER	G-CDFS
Coldplay	Bombardier CRJ-200LR Regional Jet	G-MKSA
Elvis Presley	Convair 880	N880EP 'Lisa Marie'
Elvis Presley	Lockheed JetStar	N777EP 'Hound Dog II'
Iron Maiden	Boeing 757	G-OJIB 'Ed Force One'
Iron Maiden	Boeing 757	G-STRX 'Ed Force One'
KISS	Boeing 747SP	N532PA
KISS	Bombardier CRJ-200ER Regional Jet	PH-AAG
Led Zeppelin	Boeing 720	N7201U 'The Starship'
Led Zeppelin	Boeing 720	N7224U 'Caesar's Chariot'
Pixie Lott	Boeing 757	G-ZAPU
U2	Hawker Siddeley HS-125-3B	G-MRFB
U2	Boeing 727	N511DB
U2	Boeing 727	N724CL
U2	Boeing 737	HB-IIQ
U2	Airbus A-320	C-FPWE
U2	Bae 146	G-BPNT
U2	McDonnell Douglas MD-83	F-GMLK
U2	Airbus A-320	C-GQCA
U2	McDonnell Douglas MD-83	N949NS

Elvis Presley's private Convair 880 jet 'Lisa Marie' is now on public display at his Graceland home in Memphis, Tennessee. *J.B. Curio (Flickr.com)*

✈ PTEROMERHANOPHOBIA

Notable persons who have been affected by a fear of flying:

Agnetha Fältskog (musician)
Aretha Franklin (musician)
Cher (musician)
Colin Farrell (actor)
Dalai Lama (religious leader)
Doris Day (actress/musician)
Dennis Bergkamp (footballer)
Lars von Trier (film director)
Jennifer Aniston (actress)
Jennifer Connelly (actress)
Jimmy Johnstone (footballer)
Joseph Stalin (dictator)
Kirsten Dunst (actress)
Kim Jong-Il (dictator)
Ronald Reagan (politician)
Sam Shepard (actor, played Chuck Yeager in *The Right Stuff*)

Sean Bean (actor)
Stanley Kubrick (director)
Whoopi Goldberg (actress)
Woody Allen (actor/director)

 COLD WAR CODENAMES

NATO gave reporting names to aircraft built behind the Iron Curtain. They followed specific conventions: bombers were assigned names beginning with B; fighters and ground-attack aircraft were given names beginning with F; cargo and commercial aircraft were assigned names starting with C; and helicopters started with H. M was used for miscellaneous types.

Bombers:

Backfire – Tupolev Tu-22M
Badger – Tupolev Tu-16
Beagle – Ilyushin Il-28
Bear – Tupolev Tu-95
Bison – Myasishchev M-4

Blackjack – Tupolev Tu-160
Blinder – Tupolev Tu-22
Bounder – Myasishchev M-50
Brewer – Yakovlev Yak-28

Fighters:

Fagot – MiG-15
Farmer – MiG-19
Feather – Yakovlev Yak-15/17
Fencer – Sukhoi Su-24
Fiddler – Tupolev Tu-28/128
Firebar – Yakovlev Yak-28
Firkin – Sukhoi Su-47
Fishbed – MiG-21
Fishpot – Sukhoi Su-9/11
Fitter – Sukhoi Su-7/17/20/22
Flagon – Sukhoi Su-15
Flanker – Sukhoi Su-27/30/33/35/37

Flashlight – Yakovlev Yak-25
Flogger – MiG-23 / MiG-27
Flora – Yakovlev Yak-23
Forger – Yakovlev Yak-38
Foxbat – MiG-25
Foxhound – MiG-31
Freestyle – Yakovlev Yak-141
Fresco – MiG-17
Frogfoot – Sukhoi Su-25
Fulcrum – MiG-29/33/35
Fullback – Sukhoi Su-34

Cargo and commercial aircraft:

Camber – Ilyushin Il-86
Camel – Tupolev Tu-104
Camp – Antonov An-8
Candid – Ilyushin Il-76
Careless – Tupolev Tu-154
Cash – Antonov An-28

Cat – Antonov An-10
Clank – Antonov An-30
Classic – Ilyushin Il-62
Cleat – Tupolev Tu-114
Cline – Antonov An-32
Clobber – Yakovlev Yak-42

Clod – Antonov An-14
Coaler – Antonov An-72/74
Cock – Antonov An-22
Codling – Yakovlev Yak-40
Coke – Antonov An-24
Condor – Antonov An-124

Cookpot – Tupolev Tu-124
Coot – Ilyushin Il-18
Cossack – Antonov An-225
Crusty – Tupolev Tu-134
Cub – Antonov An-12
Curl – Antonov An-26

Helicopters:
Halo – Mil Mi-26
Hare – Mil Mi-1
Harke – Mil Mi-10
Hat – Kamov Ka-10
Havoc – Mil Mi-28
Haze – Mil Mi-14
Helix – Kamov Ka-27/28/29/31/32
Hen – Kamov Ka-15
Hermit – Mil Mi-34
Hind – Mil Mi-24

Hip – Mil Mi-8/17
Hog – Kamov Ka-18
Hokum – Kamov Ka-50/52
Homer – Mil Mi-12
Hoodlum – Kamov Ka-26/126
Hook – Mil Mi-6
Hoplite – Mil Mi-2
Hormone – Kamov Ka-25
Hound – Mil Mi-4

Miscellaneous:
Madge – Beriev Be-6
Maestro – Yakovlev Yak-28
Mail – Beriev Be-12
Mainstay – Beriev A-50
Mandrake – Yakovlev Yak-25
Mangrove – Yakovlev Yak-27

Maxdome – Ilyushin Il-80
May – Ilyushin Il-38
Mermaid – Beriev A-40
Midas – Ilyushin Il-78
Mitten – Yakovlev Yak-130
Moss – Tupolev Tu-126

FAMOUS FEATS OF AVIATION: DOOLITTLE RAID

Desiring a morale boost following the surprise raid on Pearl Harbor the previous December, in early 1942 the USA decided to attack Tokyo. As there was no airfield close enough, the aircraft chosen for the raid – B-25 Mitchell medium bombers – would have to operate from an aircraft carrier. The bombers were too big to land on the deck, so they had to be loaded by crane. It also meant the raid would have to be one-way.

On the morning of 18 April, the sixteen Mitchells took off from the USS *Hornet* led by the mission's instigator: Lieutenant Colonel James Doolittle. They had to start the mission further away from the target than planned as the carrier group had been spotted by a Japanese boat and it was feared the

James Doolittle pilots the first B-25 Mitchell bomber off the deck of the aircraft carrier USS *Hornet*. Being first, and with all the other aircraft parked behind, he had the least amount of deck to take off from. *US National Archives*

surprise element was lost. As it turned out the aircraft were able to bomb Tokyo with complete surprise, the first time the Japanese mainland had been subjected to air attack. Once they had dropped their bombs, the plan was to fly on to airfields in China. None of the Mitchells made their intended landing spots and the crews crash landed or bailed out. Six crewmen died, three being executed by the Japanese, two drowned and one died of illness while a prisoner. While the raid was a propaganda success actual damage to the targets was slight. However, it had a far wider and tragic impact in China where the Japanese exacted retribution for assistance given to the American crews. It is estimated 250,000 Chinese were killed around this time.

✈ INCREDIBLE MACHINES: CAPRONI CA.60 FLYING BOAT

The Ca.60 was one of the most unusual-looking aircraft ever built. It had nine wings, eight engines and its fuselage was a converted houseboat. It weighed 55,000 pounds. Designed to fly from Italy to America, it didn't make it. In fact, on its first flight, in March 1921, it barely made it into the air before crashing back into Lake Maggiore. An Italian historian, on seeing the machine, was quoted as saying: 'It would not have looked out of place sailing up the English Channel with the Spanish Armada.' The wreckage was later burnt.

 BIGGLES BOOKS

In days gone by one fictional hero of the skies stood out: Major James Bigglesworth, more commonly known as Biggles. The pilot who started off in the First World War, flew through the Second World War, then went on to face down beastly criminals in the Special Air Police in the post-war years was a legend to millions of young readers. The intrepid chap with his chums Algy, Ginger and Bertie starred in over a hundred books, amongst them:

The essential *Biggles Learns to Fly*.
The non-dromedary *The Camels are Coming*.
The compass-driven *Biggles Flies East*, *Biggles Flies West*, *Biggles Flies South*, and *Biggles Flies North*.
The frankly improbable *Biggles Sweeps the Desert*.
The unlikely *Biggles Fails to Return*.
The recruitment-heavy *Biggles Gets His Men*.
The running out of ideas *Another Job for Biggles*.
The exhausting *No Rest for Biggles*.
The hard-pressed domestic *Biggles Makes Ends Meet*.
The serial-killing *Biggles Buries the Hatchet*.
The lottery hopeful *Biggles Forms a Syndicate*.
The hinting at a secret type of love *Orchids for Biggles*.
The spoiler *Biggles and the Plot that Failed*.
The frightfully British *Biggles Sorts it Out*.
The intriguing *Biggles Sees Too Much*.
The regressing to childhood *Biggles Does Some Homework*.
The about time too *Biggles Takes a Holiday*.

 DID YOU KNOW?

1.3 million:
The number of civil flights per each reported accident.

 AEROBATIC MANOEUVRES

Two-point Roll	Aileron Roll
Four-point Roll	Loop
Hesitation Roll	P-Loop
Barrel Roll	Q-Loop
Snap Roll	Square Loop

Humpty-Bump
Split-S
Chandelle
Immelmann
Derry Turn
Stall Turn
Cuban Eight

Half Cuban Eight
Half Reverse Cuban Eight
Lazy Eight
Tailslide
Wing Over
Spin
Pugachev's Cobra

✈ ... GIVES YOU WINGS

Red Bull founder Dietrich Mateschitz has built up a collection of aircraft which are housed in Hangar 7 in Salzburg, Austria. All the aircraft are painted with the corporate colours of the energy drink. The Flying Bulls fleet:

Alpha Jet
BD-5J Microjet
Bell 47
Bell Cobra
Bo 105
Boeing PT-17 Stearman
Cessna C 337
Cessna CE 208 Amphibian Caravan

Chance Vought F4U-4 Corsair
Douglas DC-6B
Eurocopter EC 135
Fairchild PT-19
Lockheed P-38L Lightning
North American B-25J Mitchell
North American T-28B Trojan
Pilatus Turbo Porter PC-6

INCREDIBLE MACHINES: 'SPRUCE GOOSE'

Howard Hughes is remembered for the wildly eccentric latter part of his life, when he retreated into a hermitic and paranoid lifestyle. Despite his sad demise, it shouldn't be forgotten Hughes was a dynamic presence in aviation, being involved in innovative aircraft production and airline operation. His most famous aircraft was the Hughes HK-1/H-4 'Hercules', commonly known as the 'Spruce Goose' (despite it being made of birch wood). The huge military seaplane was too late for the war but Hughes continued with its manufacture. One day in November 1947 he was on board to oversee some taxiing trials. After several runs Hughes took the aircraft around for one more. All eight throttles were pushed all the way forward. After a long 'taxi' the aircraft's gigantic wings produced enough lift and the biggest aircraft in the world took to the air. It flew for less than a mile but Hughes had proven it could fly. It went into its climate-controlled hangar and never left to fly again.

✈ DID YOU KNOW?

1,087:
The world record for the greatest number of passengers flown on one aircraft. In 1991 in the midst of a civil war in Ethiopia the Israelis evacuated 14,500 Jews and flew them to Israel. They were flown out of the African country in El Al aircraft including the 1,087 in a Boeing 747. Seats that were designed for four passengers took six. Even this huge number of passengers is thought to be an underestimate, as some children hid under their mothers' robes.

✈ GREAT PLANES: CONCORDE

For over thirty years there was one aircraft that merited a change in grammar. When one was sighted people said 'There's Concorde' and not 'There's *a* Concorde'. It was a special aircraft, a long-distance airliner that cruised at twice the speed of sound at 60,000ft, allowing passengers to see the curvature of the planet beneath. Due to heat caused by air friction when supersonic the airframe expanded by up to 8in and a number of the surviving aircraft have flight engineers' caps stuck forever in the gap between the engineer's instrument panel and the cockpit bulkhead. It flew so fast it was

Concorde G-BOAF lands at the 1998 Farnborough Airshow. In November 2003 it made the last ever Concorde flight when it was flown to Filton, near Bristol. Its arrival was watched by thousands of people. *Derek N. Ferguson*

possible to see a sunrise in the west, the aircraft flying faster than the Earth's rotation. In 1974, in a demonstration of its capabilities, an Air France Concorde took off from Boston, USA, as a Boeing 747 took off from Paris, heading to the New England destination. Concorde made it to France, refuelled, took off and arrived back at Boston before the jumbo jet could land!

Concorde had flown fatality-free for thirty years but in 2000, an Air France Concorde crashed minutes after taking off from Paris' Charles de Gaulle Airport. Concorde returned to service after structural changes but it was not to last long. In 2003 it was announced it would be retired. It remains one of the few aircraft types to be retired with no replacement in the wings.

FAMOUS FEATS OF AVIATION: RUTAN ALL OVER THE WORLD

Several non-stop flights have been made around the world but until 1986 one hadn't been achieved with the aircraft remaining un-refuelled during its flight. The innovative designer Burt Rutan came up with a unique machine named *Voyager* for this purpose. It was very thin, made up of one slender fuselage and two parallel booms, which were used to carry fuel. Its fragile-looking wings spread out to a span of 110ft. On 14 December Burt's brother Dick Rutan and co-pilot Jeana Yeager took off from California. On their epic journey they both suffered from extreme fatigue and also hypoxia, due to having to fly higher to avoid bad weather, but nine days after taking off they completed their record-breaking voyage and landed at Edwards Air Force Base. Fifty thousand people gathered to see them arrive back. They had flown 26,358 miles at an average speed of 116mph. Their flight was more than double the existing distance record for an un-refuelled aircraft.

✈ BARNSTORMING

After the First World War there was a large surplus of unwanted aircraft and pilots. Many of the machines were sold off cheaply, particularly the Curtiss Jenny biplane. Looking to earn a living through flying, pilots took to the air as 'barnstormers'. They would turn up at a town and offer pleasure flights to the locals and then put on a special display. With competition fierce the aviators would try more and more outlandish aerial stunts to entice the audience. Most of them were performed without parachutes:

Looping-the-loop – wing-walking – playing tennis on the upper wings – changing the aircraft's tyre – climbing down a rope ladder to a speeding car – clambering from a car up on to a wing – archery – transferring from one plane to another – dancing the Charleston – being lowered in a strait-jacket then climbing back once free – being dragged along the ground from the undercarriage as the plane lands – hanging under the plane by a strap held between the teeth – flying into a barn.

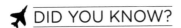

DID YOU KNOW?

2.44 billion:
The number of airline passengers in 2010, worldwide.

US AIR FORCE THEME

'Off we go into the wild blue yonder,
Climbing high into the sun;
Here they come zooming to meet our thunder,
At 'em boys, Give 'er the gun! (Give 'er the gun now!)
Down we dive, spouting our flame from under,
Off with one helluva roar!
We live in fame or go down in flame. Hey!
Nothing'll stop the US Air Force!'

AIR-TO-AIR MISSILES

AMRAAM – Advanced Medium Range Air-to-Air Missile
ASRAAM – Advanced Short Range Air-to-Air Missile
BVRAAM – Beyond Visual Range Air-to-Air Missile
LRAAM – Long-Range Air-to-Air Missile
MRAAM – Medium-Range Air-to-Air Missile
SRAAM – Short-Range Air-to-Air Missile
WVRAAM – Within Visual Range Air-to-Air Missile

In this image candy can be seen dropping from this Douglas C-54 transporter aircraft during the Berlin Airlift. Three hundred C-54s were used during the huge operation to supply the German capital. *USAF*

✈ OPERATION 'LITTLE VITTLES'

The Berlin Airlift was the biggest supply mission in aviation history. Over 2.3 million tons of supplies were flown into isolated West Berlin following the closing of the normal access routes by the Soviet Union, who were keen to see the western forces leave the encircled city. The Allies decided to form an 'air bridge' utilising thousands of cargo aircraft to supply the 2 million inhabitants. The United States called their part 'Operation Vittles'. One of the American pilots, Lieutenant Gail S. Halvorsen, had started dropping sweets by parachute for German children. His popularity earned him the nicknames 'The Chocolate Flyer' and 'Uncle Wiggly Wings'. Other pilots became involved in 'Operation Little Vittles' and around 23 tons were dropped by the 'Candy Bombers' by the end of the blockade.

ALLITERATIVE AIRCRAFT APPELLATIONS

Some British aircraft manufacturers had a preference for naming their products alliteratively with the same letter as the company name. One of these was Vickers:

Vagabond	Vespa
Valentia	Viastra
Valetta	Victoria
Valiant	Viget
Valparaiso	Viking
Vampire	Vildebeest
Vanguard	Vimy
Vanox	Vincent
Varsity	Vireo
Vellore	Virginia
Vellox	Viscount
Vendace	Vivid
Venom	Vixen
Vernon	Vulcan

 BRAVERY IN THE AIR: JAMES NICOLSON

Only one Victoria Cross was awarded to the RAF's Fighter Command during the whole of the Second World War. The citation for the medal stated:

'Flight Lieutenant James Brindley Nicolson. (39329) – No 249 Squadron.
During an engagement with the enemy near Southampton on 16th August 1940, Flight Lieutenant Nicolson's aircraft was hit by four cannon shells, two of which wounded him whilst another set fire to the gravity tank. When about to abandon his aircraft owing to flames in the cockpit he sighted an enemy fighter. This he attacked and shot down, although as a result of staying in his burning aircraft he sustained serious burns to his hands, face, neck and legs.'

Nicolson recovered from his injuries and resumed flying. He was posted to India where he flew Beaufighters, although it was as an observer in a B-24 Liberator that he was killed only three months before the end of the war.

✈ SEX, LIE AND VID DEO TAP

Each airport has a three-letter unique code assigned to it by the International Air Transport Association. There are over 9,000 registered. There is only room to show some:

AGE	–	Flugplatz, Germany
AND	–	Anderson, South Carolina, USA
BAP	–	Baibara, Papua New Guinea
BAW	–	Biawonque, Gabon
BIB	–	Baidoa, Somalia
BUS	–	Batumi, Georgia
DEO	–	Hyatt Regency Heliport, Dearborn, USA
DOG	–	Dongola, Sudan
DOH	–	Doha, Qatar
ELF	–	El Fasher, Sudan
FIG	–	Fria, Guinea
GUM	–	Guam International, Guam
GUT	–	Gütersloh, Germany
HAM	–	Hamburg, Germany
HUG	–	Huehuetenango, Guatemala
KAK	–	Kar, Papua New Guinea
LAV	–	Lalomalava, Samoa
LIE	–	Libenge, Democratic Republic of the Congo
MAD	–	Barajas, Spain
NOD	–	Norden, Germany
NOO	–	Naoro, Papua New Guinea
NUT	–	Nutuve, Papua New Guinea
POM	–	Jackson Field, Papua New Guinea
POW	–	Portoroz, Slovenia
PUS	–	Gimhae, South Korea
RAT	–	Raduzhnyi, Russia
SEX	–	Sembach, Germany
TAP	–	Tapachula International, Mexico
TAT	–	Poprad/Tatry, Slovakia
TUT	–	Tautu, Papua New Guinea
VEG	–	Maikwak, Guyana
VID	–	Vidin, Bulgaria

WORDS OF WISDOM

'If God had intended us to fly, He would never have given us the railways.'

Michael Flanders, entertainer.

LAST WORDS

Cockpit voice recorders were installed in aircraft to provide essential input to any accident investigation. In amongst the technical detail of these incidents they can give a powerful insight into the last moments of those on board:

'Nose up ... Nose up ... Power.'
Captain Takahama. Japan Airlines, Flight 123. Mechanical failure saw catastrophic damage to the tail fin. The Boeing 747 crashed into a mountain. 520 people were killed, in the worst single-plane air accident.

First Officer: 'Larry, we're going down, Larry ...'
Captain: 'I know it.'
First Officer Roger Pettit and Captain Larry Wheaton were on Air Florida Flight 90 when the Boeing 737 took off from Washington National Airport without being properly de-iced. Ice on the wings prevented the aircraft from attaining a good climb-out rate. It crashed into the Potomac River. Seventy-eight died.

'Put out your cigarette, this is an emergency descent. Put the mask over your nose and mouth and adjust the headband. Put the mask over your nose and mouth and adjust ...'
Public address recording, Korean Air Flight 007. The Boeing 747 was heading to Seoul when it drifted off its intended course and flew into Soviet airspace. A Soviet air defence fighter fired two missiles which exploded and caused the airliner to lose cabin pressure. It lost height and crashed into the sea, resulting in all 269 on board being killed.

'Oh ... this can't be.'
Captain Arturo Valdes Prom. An Aeromexico McDonnell Douglas DC-9 operating as Flight 498 collided with a light aircraft over Cerritos in California. All on board both aircraft were killed and fifteen on the ground also died.

'Ma I love you.'

An unidentified voice on the cockpit voice recorder on board Pacific Southwest Airlines Flight 182 after it collided with a Cessna 172 light aircraft over San Diego in California. The Boeing 727 caught fire and quickly departed from controlled flight. All occupants of both aircraft died as well as another seven on the ground.

'Pull up. That's it, I'm dead.'

The flight engineer on board Surinam Airways Flight PY764. The DC-8 was attempting to land at Paramaribo-Zanderij in Suriname using unreliable equipment when it impacted the ground. Only eleven of those on board survived.

'Negative, we're trying Le Bourget.'

First Officer, Air France Concorde Flight 4590. The Concorde suffered catastrophic damage after a piece of metal caused a fire when taking off from Charles de Gaulle Airport. The aircraft struggled into the air and the crew intended to make an emergency landing at nearby Le Bourget Airport. The aircraft stalled and crashed into a hotel. All 109 on board and four on the ground were killed.

✈ WORDS OF WISDOM

'The gratitude of every home in our Island, in our Empire, and indeed throughout the world, except in the abodes of the guilty, goes out to the British airmen who, undaunted by odds, unwearied in their constant challenge and mortal danger, are turning the tide of the world war by their prowess and by their devotion. Never in the field of human conflict was so much owed by so many to so few.

'All hearts go out to the fighter pilots, whose brilliant actions we see with our own eyes day after day; but we must never forget that all the time, night after night, month after month, our bomber squadrons travel far into Germany, find their targets in the darkness by the highest navigational skill, aim their attacks, often under the heaviest fire, often with serious loss, with deliberate careful discrimination, and inflict shattering blows upon the whole of the technical and war-making structure of the Nazi power. On no part of the Royal Air Force does the weight of the war fall more heavily than on the daylight bombers who will play an invaluable part in the case of invasion and whose unflinching zeal it has been necessary in the meanwhile on numerous occasions to restrain.'

Winston Churchill, 20 November 1940.

✈ INCREDIBLE MACHINES: EKRANOPLAN

The MD-160 Ekranoplan (Wingship) was an aircraft like no other. This was because it wasn't strictly an aircraft. The giant Soviet-built machine had wings and jet engines and a fuselage but was not intended to fly high above the Earth as an aircraft might be expected to do. Instead it was to skim only a few metres above the water utilising ground effect, i.e. the increased lift afforded by flying just above the surface. The Ekranoplan needed it as it weighed over 280 tons and was 240ft in length, making it one of the largest machines ever to take to the air. Although very big, the mighty MD-160 was actually a smaller version of the huge ten-engined 'Caspian Sea Monster' that was first seen in the 1960s. The MD-160, which was equipped with six anti-ship cruise missile launchers sitting on top of the fuselage, saw service with the Russian Navy for over a decade until being retired and left in dry dock in the 1990s.

✈ MULTI-ENGINED AIRCRAFT

Nowadays the trend is to have fewer engines, such has been the development of technology, so transatlantic airliners are designed with two when years ago they would have had four. These are some of the aircraft that required even more than four:

No of engines	Engine Type	Aircraft Name	Type	First Flight
12	Curtiss Conqueror piston engines	Dornier Do X	flying boat	1929
10	Pratt & Whitney Wasp Major radial engines (x6), General Electric J47 turbojets (x4)	Convair B-36 Peacemaker	bomber	8 August 1946
"	Bristol Proteus turboprops	Saunders-Roe Princess	long-range civil flying boat	22 August 1952
9	Pratt and Whitney TF106 turbofan for forward thrust, Rolls-Royce RB162 turbofans (x8) for vertical lift	Dassault Mirage IIIV	vertical take-off supersonic fighter	12 February 1965

8	Kuznetsov NK-87 turbofans	MD-160 Ekranoplan	seaplane	Unknown
"	Liberty L-12 V-12 piston engines	Caproni Ca-60	flying boat	4 March 1921
"	Mikulin AM-34FRN V-12 piston engines	Tupolev ANT-20 Maxim Gorky	propaganda aircraft	1934
"	Allison J35-A-5 axial flow turbojets	Northrop YB-49	flying wing bomber	21 October 1947
"	Pratt & Whitney Wasp Major radial engines	Hughes HK-1/ H-4 Hercules	seaplane	2 November 1947
"	Bristol Centaurus radial engines	Bristol Brabazon	airliner	4 September 1949
"	Pratt & Whitney TF33-P-3/103 turbofans (H-model)	Boeing B-52 Stratofortress	bomber	15 April 1952
"	Pratt & Whitney J57-P-3 turbojets	Convair YB-60	bomber prototype	18 April 1952
7	Mikulin AM-34F V-12 piston engines	Kalinin K-7	bomber/ airliner	11 August 1933
6	Liberty L-12A piston engines	Witteman-Lewis NBL-1 Barling Bomber	bomber prototype	22 August 1923
"	Isotta-Fraschini Asso piston engines	Caproni Ca-90	bomber prototype	1929
"	Napier Lion piston engines	Tarrant Tabor	bomber	26 May 1919 (crashed before it could take off)
"	Mikulin AM-34 V-12 piston engines	Tupolev ANT-16	bomber prototype	3 July 1933
"	Mikulin M-34R piston engines	Tupolev ANT-22	flying boat	8 August 1934

"	Hispano-Suiza 12N piston engines	Latécoère 521 Lieutenant de Vaisseau Paris	flying boat	10 January 1935
"	Jumo 207C inline diesel piston engines	Blohm & Voss BV 222	flying boat	7 September 1940
"	Wright R-2600-A5B Cyclone radial piston engines	Latécoère L 631	flying boat	4 November 1942
"	Gnome-Rhone radial piston engines	Messerschmitt Me 323 Gigant	transport aircraft	1942
"	Daimler-Benz DB 603G inline piston engines	Blohm & Voss BV 238	flying boat	1944
"	BMW 801 radial piston engines	Junkers Ju 390	bomber/ maritime patrol/ transport	20 October 1943
"	Pratt & Whitney Wasp Major radial piston engines	Convair XC-99	cargo aircraft	23 November 1947
"	Pratt & Whitney Wasp Major radial piston engines (x4) and General Electric J47 turbojets (x2)	Boeing KB-50J/K	tanker	1955
"	Pratt & Whitney Wasp Major radial piston engines (x4) and General Electric J47 turbojets (x2)	Boeing KC-97L Stratotanker	tanker	1964
"	General Electric J47 turbojets	Boeing B-47 Stratojet	bomber	17 December 1947
"	General Electric J35 turbojets	Martin XB-48	prototype bomber	22 June 1947

"	Rolls-Royce Griffon piston engines (x4) and Rolls-Royce Viper 203 turbojets (x2)	Avro Shackleton MR.3 Phase 2	maritime patrol	Unknown
"	Ivchenko Progress D-18T turbofans	Antonov An-225 'Mriya'	transport	21 December 1988
"	General Electric YJ93 turbojets	North American XB-70 Valkyrie	prototype bomber	21 September 1964
5	Junkers Jumo 211 piston engines	Heinkel 111Z 'Zwilling'	glider tug	1941

This scale model of the Dornier Do-X is on display at the Fliegermuseum Altenrhein in Switzerland. The huge seaplanes were built at Altenrhein, which sits on the edge of Lake Constance. *Kecko (Flickr.com)*

 A IS FOR ...

The standard international phonetic alphabet is used for communication purposes to signify each individual letter. So A is Alpha, B is Bravo, C is Charlie and so on. However, in 1921 a different set of assigned words was used for radio-telephonic communications:

A – Ac   C – Charlie
B – Beer   D – Don

E – Edward
F – Freddie
G – George
H – Harry
I – Ink
J – Johnnie
K – King
L – London
M – Monkey
N – Nuts
O – Orange

P – Pip
Q – Queen
R – Robert
S – Sugar
T – Too
U – Uncle
V – Vic
W – William
X – X-ray
Y – Yorker
Z – Zebra

 DON'T WALK UNDER LADDERS

A few superstitions of airline passengers flying from Detroit, 2011:

Lifting feet off the cabin floor during take-off.
Rolling fingers to imitate propellers.
Raising seat armrest at moment of lift-off.
Buying crossword puzzle books.
Holding on to in-flight magazine.
Repeating phrase 'flaps and slats'.
Not saying goodbye to wife.
Counting to 15.
On boarding patting the outside of the aircraft.
Carrying elephant charm bracelet.
Taking M&M sweets.
Stepping on to aircraft with the right foot.
Listening to Bonnie Raitt.

NO FLY ZONE

The airlines that cancelled orders for Concorde:

Air Canada
Air India
American Airlines
Braniff
CAAC
Continental

Eastern
Iran Air
Japan Air Lines
Lufthansa
Middle East Airlines – Air Liban
Pan American

Qantas   TWA
Sabena   United Airlines

✈ PHANTOM DIVE

The Gloster Meteor was the first jet fighter to enter British squadron service. The airbrakes were located between the wings and the fuselage. Every Meteor pilot had it drummed into him never to lower the undercarriage when the speed brakes were out as it would cause the wing to lose lift, and the aircraft would lose its ability to maintain controlled flight. This phenomenon was known as a 'Phantom Dive'. The only recovery was to retract the undercarriage, pull the airbrakes in and pull up from the dive. That only worked if there was enough altitude. Pilots in the landing circuit rarely had enough height nor time to effect a recovery. Ejector seats were not fitted to Meteors and losses were appalling with over 850 being lost. In 1988 the RAF lost a vintage Meteor when it crashed at an airshow at Coventry. The airbrakes were seen to be extended as it came into land.

MYSTERIES OF THE SKY: AMY JOHNSON

Amy Johnson was Britain's premier aviatrix, a darling of the Golden Age of Aviation. In 1930 she achieved a notable success, becoming the first woman to fly solo from Britain to Australia. Johnson's achievement was remarkable considering her only previous long-distance flight was from London to her hometown Hull. She was fêted after her flight and received two high-profile honours – the CBE and being modelled in Madame Tussaud's.

Johnson made other notable flights, including flying from London to Moscow in one day. In 1940 she joined the Air Transport Auxiliary. Unfortunately she achieved a tragic first, becoming the first ATA pilot to be killed in the war. She was delivering an Airspeed Oxford twin-engined utility aircraft from Prestwick to Oxfordshire when she was seen to bail out over the Thames Estuary. She was 100 miles off course and at the end of the time the plane's fuel would last. A Royal Navy trawler HMS *Haslemere* went to her aid and in strong seas and snowy conditions the captain dived in to help. He later died of hypothermia. Johnson's body was never discovered and a host of theories have arisen to explain what happened. Was she on a mission to deliver a spy to Europe or was she shot down by British anti-aircraft guns?

Could it have been German fighters? One report claimed that the trawler, in attempting to manoeuvre closer, ran into her with its propellers. Whatever the reason, Britain had lost a national heroine.

✈ AERIAL COMMANDO BOOKS

Much pocket money was, and continues to be, spent on the small-format comic books printed by DC Thomson. Of the thousand published these are some of the aviation-themed editions:

A Stirling Called Satan
Ace Without Honour
Battle Squadron
Black Ace
Bombs on Target!
Boss of the Sky
Brigand Squadron
Castaway Squadron
Close Support!
Convict Squadron
Danger Below
Daredevil DFC
Death of a Wimpey
Deck-Level Dawson
Desert Dogfight
Dragon Pilot
Duel over the Desert
Eyes in the Night
Fearless Foster
Fearless Freddy
Fighting Eagle
Fly Fast – Shoot Fast
Fly Into Fear
Flying Fury
Flying Phantom
Flying Rivals
Flying Viper
Free French Ace
Fury Strike

Ghost in the Cockpit
Ghost Pilot
Giant Duel
Giant!
Hit 'Em Hard!
Hoodoo Ace
Hunger for Glory
Jack's Private War
Kill the Fuhrer
Messerschmitt Storm
Mongrel Squadron
Pilot for Hire
Sky Sniper
Spitfire Spirit
Spy in the Sky
Stormbird Strike
Swastika Squadron
Target – China!
The Convoy-Killer
The Silver Spitfire
Too Late for Glory?
Transport Ace
Upside-Down Ace
Valley of Death
Walrus Patrol
War of the Tin Goose
Wings of the Warrior
Wings of War

TSR-2

'All modern aircraft have four dimensions: span, length, height and politics. TSR-2 simply got the first three right.'

Sir Sydney Camm, chief designer, Hawker.

Of all the cancelled aircraft projects in British aviation history – and there have been more than a few – none causes as much anguish as the TSR-2. Designed as a fast reconnaissance and ground attack, nuclear-capable bomber it first flew in 1964. Built by BAC, its two crew were carried in a distinctive streamlined airframe. The specification promised much: it was to fly at Mach 2.35 at 40,000ft and be able to fly faster than the speed of sound at sea level. However its development encountered delays over the engines, and other technical issues. But as Sydney Camm remarked it was the politics that were to ground it. It was expensive – Britain's biggest-ever aircraft project – and the Labour government were desperate to seize on any possible budget savings. In April 1965 it was announced that TSR-2 was to be cancelled and replaced by American-built F-111s. They never arrived. The F-111 programme was hit by delays and cost overruns and the RAF was forced to take aircraft from an unlikely source – the Royal Navy. Two TSR-2s survive in British museums. The only one to fly was used for target practice.

LUNA HABITABILIS

Thomas Gray wrote a poem in 1737 which contained these lines, which some have taken as prescient of the Battle of Britain:

'When thou shalt lift thine eyes
To watch a long drawn battle in the skies,
While aged peasants, too amazed for words,
Stare at the flying fleets of wond'rous birds:
England, so long the mistress of the sea,
Where wind and waves confess her sovereignty,
Her ancient triumphs yet on high shall bear
And reign, the sovereign of the conquered air.'

A Typhoon in RAF service with 6 Squadron. The Eurofighter 2000 was named Typhoon in 1998. The multi-role aircraft is used in both air defence and ground attack roles. *Norman H. Ferguson*

✈ AVIATOR ALIASES

Ace – Aeronaut – Aeroplanist – Airman – Aviatrix – Navigator of the Air:

Terms for pilots in the early years of aviation. Only airman remains in common usage.

✈ EUROFIGHTER 2000

According to publicity material, the principal features of the Eurofighter 2000 are:

Advanced aerodynamic layout combining a delta wing and foreplane canards; new construction techniques and extensive use of advanced materials; Active Control Technology for optimum manoeuvrability; highly integrated avionic systems; highly automated cockpit with Helmet Mounted Symbology and Direct Voice Input; multi-mode, pulse-Doppler radar with multi-target capability; comprehensive, internally housed defensive aids; heavy and varied weapons load.

BROCK, POMEROY AND BUCKINGHAM

Not a legal firm but types of ammunition used to combat the 'Zeppelin menace' in the First World War. In January 1915 the war had been brought to the home front with the first airship attack on Britain. The dirigibles caused alarm amongst the civilian population, who had thought themselves relatively safe. It took over a year and a half until British fighters were able to counter this aerial threat when German airship SL 11 was shot down by a Royal Flying Corps Be2c biplane flown by Lieutenant W. Leefe Robinson in September 1916. Robinson was awarded a Victoria Cross for his endeavours. What made the difference was the ammunition now available to be fired at the hydrogen-filled airships. Brock and Pomeroy explosive bullets went into machine-gun magazines alongside the phosphorus-containing Buckingham and 'Sparklets', which contained magnesium. Zeppelins were now much easier targets and were forced to fly higher to avoid the fighters. This wasn't always successful and one was lost to an aircraft piloted by Egbert Cadbury (from the chocolate-making family) in August 1918. The dirigibles, whose main effectiveness was psychological, were phased out of operations.

 ## DID YOU KNOW?

9.232 miles (minus 150ft):
The distance of the world's longest golf putt, made in mid-air in 1999. The European Ryder Cup team were flying over the Atlantic to Massachusetts in Concorde. Spanish golfer Jose Maria Olazabal made the putt, which travelled 150ft down the length of the airliner's cabin. The putt lasted 26.17 seconds, and with them travelling at Mach 2, the distance travelled was 9.232 miles!

 ## REAL AMERICAN HEROES

A few tough-sounding American aircraft names:

Avenger	Fighting Falcon
Banshee	Flying Fortress
Bronco	Liberator
Demon	Marauder
Destroyer	Mauler
Devastator	Mustang
Eagle	Nighthawk
Fireball	Prowler

Sabre
Savage
Skywarrior
Stratofortress
Starfighter
Super Savage

Thunderbolt
Tigercat
Tigershark
Valkyrie
Vigilante

 # GREAT PLANES: BOEING 747

When the Boeing 747 first flew in 1969 it was intended to be the prime long-haul airliner. It succeeded. The four-engined 'Jumbo Jet' has been in production for over forty years, with a new variant, the 747-8, announced in 2005. The 747 has seen service with eighty airlines, including major carriers such as British Airways, Lufthansa, Air France, Pan Am and Japan Airlines. It was the first 'wide-body' passenger jet, having two aisles and the first airliner to have two levels, the first-class passengers being seated on the top deck behind the cockpit area. An indication of the type's success is that 3.5 billion passengers have flown on 747s. Over 1,400 747s have been manufactured, including two used as the US President's 'Air Force One' and a special version adapted to carry NASA's Space Shuttle.

An Air China Cargo Boeing 747 taxiing at Manchester Airport. The aircraft was originally intended to be a transport aircraft, hence the positioning of the cockpit above the main fuselage. *Derek N. Ferguson*

✈ MYSTERIES OF THE SKY: AMELIA EARHART

Amelia Earhart is perhaps the most famous female pilot. The American was the first woman to fly the Atlantic and then became the first to fly it alone. She attained other notable aviation successes and it was while attempting another that she met her end. In July 1937 Earhart and her navigator Fred Noonan were on the final stages of completing a circumnavigation flight. They were due to arrive at Howland Island, a small island in the Pacific. Showing how well regarded the aviatrix was, the airport had been specially built for her by the government. She was never to reach it. A US Coast Guard ship had been positioned close to the island to communicate with the Lockheed Electra Model 10E as it approached from Lae in Papua New Guinea. Despite being able to hear Earhart clearly on the radio, the sailors were unable to make proper contact with her. The plane flew on and missed its intended landing site. Its disappearance caused an air and sea search but nothing was found. Radio messages from Earhart were picked up in the subsequent days as far away as the continental USA but some were regarded as hoaxes. In 1940 part of a human jawbone, a woman's shoe and a sextant box were found on an uninhabited island hundreds of miles south east of Howland. Had the Lockheed plane been able to land and its crew survive only to succumb after desperately making radio transmissions? The real answer may never be known.

✈ GROOM LAKE

Groom Lake, also known as the Container, the Box, the Ranch, Dreamland, Homey Airport or simply Area 51, is one of the most talked about airfields in the world. Speculation knows no bounds when it comes to what might have flown from it, be it ultra-secret aircraft that might never have existed, or even UFOs. The documented types are interesting enough in themselves. The U-2 made its first flight from the Nevada base. The A-12, forerunner of the SR-71 Blackbird, set many high speed and altitude records there. The stealth 'fighter', the F-117 Nighthawk, was tested in extremely high levels of security, so much so that when the aircraft was revealed to the world in 1988 no one outside the project had even seen it.

One of the other secret projects at Groom Lake was the flying of Soviet-built fighters in mock dogfights against US pilots, to familiarise them with aircraft they might face if the Cold War heated up. Types that are known include the MiG-17 Fresco, MiG-21 Fishbed and the swing-wing MiG-23 Flogger.

What aircraft currently fly from the area is unknown.

F-117 Nighthawk was the world's first operational stealth aircraft. Its unconventional appearance was a result of careful design to reduce its size when picked up by hostile air defence radars. *USAF*

 THE DAM BUSTERS

Gibson, Pulford, Taerum, Hutchison, Spafford, Deering, Trevor-Roper:

These names have gone down in history as the crew members of Lancaster bomber AJ-G on the night of 16/17 May 1943, the night of one of the most remarkable and inspiring events of the Second World War: the Dams Raid. Nineteen Lancasters took off from Lincolnshire's RAF Scampton to bomb dams in Germany's Ruhr district. The bombers used a specially designed weapon: Barnes Wallis' 'bouncing bomb'. 617 Squadron was created specifically for this mission and was led by the uncompromising and not always popular Wing Commander Guy Gibson. Gibson's commitment to the war effort was total: he had flown three operational tours by the time he arrived at Scampton, including one as a night fighter pilot.

At the Möhne dam Gibson's Lancaster made the first attack, at just 60ft. It then circled to draw flak from other Lancasters as they made their bombing runs. When the dam was breached Gibson led bombers still armed with their bombs to the Eder dam. There his aircraft again acted to distract the German anti-aircraft guns. All of Gibson's crew were awarded medals with

Gibson receiving the Victoria Cross. None of the seven survived the war. Gibson was killed in 1944 over the Netherlands while on a bombing mission. He was 26.

✈ TICKET TO RIDE

The aircraft whose engines are heard at the beginning of The Beatles' song 'Back in the USSR' is a turboprop-powered Vickers Viscount.

✈ THE ORIGINS OF PLANE-SPOTTING

'Under the special conditions of modern warfare interest in the identity of aircraft is not confined to members of HM Defence Services. A desire to be able to distinguish with certainty between friendly and hostile planes is widespread and natural.'

R.A. Saville-Sneath, author of *Aircraft Recognition* (1941).

There has been a long-standing interest in plane-spotting in Britain, which can be put down to a number of things. In the years following the Wright Brothers' first flight aviation became the fad of the day. The public became fascinated by this brand new invention and were enthralled by the daring feats of the aviators. First flights were happening all the time and records were constantly being broken as planes went further and faster. Blériot, Cody, Alcock and Brown, Lindbergh, Johnson, Cobham – all received great public acclaim for their aerial endeavours.

The First World War had seen enemy aircraft such as Zeppelins and Gotha bombers flying over British soil and dropping bombs. The powers-that-be thought it useful to have an idea of what was above you and so in 1925 the first official spotters group was founded. The previous year a group of civilians, under the leadership of an army Major-General, got together in the back of a post office and set up a plotting system for keeping track of aircraft over the south east of England. Their first 'spot' was a Sopwith Snipe biplane seen over Kent. They were soon granted official recognition and in October 1925 the Observer Corps was brought into being by the grandly named Committee of Imperial Defence.

With bombers being produced with longer ranges and the resultant threat to more than just the south-east of England, more groups were set up around the country. However, there was one flaw in this defence set-up. The observers didn't identify the aircraft, they just worked out which were hostile by comparing the track of the aircraft spotted against that of any friendly aircraft launched. It's a long way from being a foolproof system and when the Second World War began there was an urgent necessity for action. Showing the keenness and ingenuity of the aviation enthusiast, groups of Corps members devised their own aircraft recognition tests and identification aids, which were soon adopted by other groups. These 'Hearkers' were the first proper plane-spotters. (The term hearker comes from hearkening, or listening, and although some observers could identify aircraft through sound, most relied on visual identification.)

One of the useful publications produced was by Observer Corps Hearker instructor R.A. Saville-Sneath. In 1941 Penguin published his famous and recently reissued *Aircraft Recognition*, volume 1, 'illustrated with official silhouettes and photographs'. It became a best-seller and had international spin-offs. A version was produced for the American market with a change of title. *What's That Plane?* sold over 350,000 copies.

Aviation enthusiasts photograph a Skyraider at an airshow at RAF Leuchars. Some airshows organise special packages allowing enthusiasts to gain better access to aircraft. *Norman H. Ferguson*

Following the Battle of Britain, in recognition of their sterling work the Corps received Royal recognition by becoming the Royal Observer Corps (ROC). As well as being of vital importance during the battle the Observer Corps had proved their ability by tracking the flight of Rudolf Hess' Messerschmitt 110 as he flew on his bizarre mission to Scotland. Despite some scoffing by the RAF as to their identification skills, they were proved correct by the wreckage of his plane. There was further kudos being dished out when it was a ROC officer that could identify the pilot as being Hess, and not the Hauptmann Horn he claimed. Following the war the ROC was wound up, later to be resurrected in order to spot not planes but mushroom clouds, as part of the civil defence network.

With a culture of identifying flying machines prevalent – and most schoolboys able to identify a Spitfire and Lancaster and the like – it was natural that spotting continued in the post-war years. The Fabulous Fifties helped sustain interest, as the British aircraft industry produced innovative type after type. The Lightning, the Brabazon, the Saunders Roe Princess, the SR.53, the Bristol 188 and the V-bombers were just some of the fantastic flying machines that could be seen around the country or at the Farnborough Airshow.

Unfortunately the industry was on the verge of collapse as government decisions and other factors led to a decline, but there was still plenty to see. Plane-spotting was well established. It continues to this day with thousands of people at the end of runways all over world enjoying a hobby that was once a much more serious undertaking.

✈ GREAT PLANES: F-4 PHANTOM

The legendary Phantom first flew in 1958. The twin-engined jet could fly at twice the speed of sound and in 1959 set an altitude record when it zoom-climbed to over 98,000ft. It was designed to be an interceptor suitable for use by the US Navy. It had an unusual design in that the outer parts of the wings turned up and the tailplanes turned down. It was nicknamed 'Double Ugly' by some. Despite its unconventional looks it proved so capable that the US Air Force and the US Marines bought it, the first time a single type had been flown by all three US military air arms.

Captain James Lovell, who later became known for his flights to the moon with Apollo 8 and Apollo 13, served in the US Navy and worked as a programme manager in flight testing and evaluating the F-4. He described it as 'the finest airplane I flew during my naval aviation career', marking out its stability characteristics in carrier operations.

This F-4 Phantom was operated by the RAF's 43 Squadron. The squadron inherited its aircraft from the Royal Navy after they were withdrawn from Fleet Air Arm use. They were replaced by the Panavia Tornado F.3. *Major Dennis A. Guyitt, USAF*

Phantoms were involved in conflicts such as the Vietnam War, the Arab-Israeli war of 1973, the Iran-Iraq conflict and the first Gulf War in 1991. The RAF sent Phantoms to the Falkland Islands to act as air defenders following the retaking of the islands in 1982.

Over 5,000 Phantoms were built and while most were operated by the American armed forces, over a thousand were exported to eleven countries around the world: Australia, Egypt, Germany, Greece, Iran, Israel, Japan, South Korea, Spain, Turkey and the United Kingdom. Although it is over fifty years since its first flight, the Phantom remains in operational service with several air forces.

✈ LUCKY ESCAPES: HARRY HAWKER

Harry Hawker was an aviation pioneer. He became chief test pilot at Sopwith, famous for building the First World War biplanes the Camel and Pup. In 1919 Hawker and Kenneth McKenzie-Grieve made an attempt to win the *Daily Mail*'s £10,000 prize for the first crossing of the Atlantic within seventy-two hours. They were forced to ditch and the prize went to Alcock and Brown. Apart from this lucky escape in the Atlantic he had also cheated death in 1914.

Hawker was known for giving looping demonstrations, able to perform them with the engine on and off. He could also loop-the-loop twelve times in succession. During one of his flights in a Sopwith Scout he carried out a loop – with engine off – at around 1,200ft. The aircraft then went into a vertical spinning dive. Eyewitnesses described it as 'like a leaf falling'. The Scout crashed into a tree and then fell to the ground. Onlookers rushed to help fearing the worst. Hawker was found standing amongst the wreckage. He was unhurt.

Harry Hawker co-founded the H.G. Hawker Engineering Company with Thomas Sopwith but was killed in an air accident in 1921.

✈ BRITISH AIRFIELD GATE GUARDS

Various military establishments in the UK have retired aircraft symbolically guarding their gates:

AAC Middle Wallop	Scout AH.1
AAC Middle Wallop	Sioux AH.1
AAC Middle Wallop	Auster AOP.9
RAF Boulmer	Phantom FGR.2
RAF Coningsby	Tornado F.3
RAF Cranwell	Jet Provost
RAF Halton	Hunter F.6
RAF Halton	Tornado GR.1
RAF Henlow	Hunter F.1
RAF Leuchars	Phantom FG.1
RAF Leuchars	Tornado F.3
RAF Linton-on-Ouse	Jet Provost
RAF Lossiemouth	Tornado GR.1
RAF Marham	Tornado GR.1
RAF Mount Pleasant	Phantom FGR.2
RAF Shawbury	Wessex HC.2
RAF St Mawgan	Shackleton
RAF Valley	Hunter T.8
RAF Waddington	Vulcan
RAF Wittering	Harrier GR.3
RAF Wyton	Canberra PR.9
RNAS Yeovilton	Sea Harrier FA.2

(AAC – Army Air Corps, RAF – Royal Air Force, RNAS – Royal Navy Air Station)

DID YOU KNOW?

1 hour, 54 minutes, 56.4 seconds:

The length of time Lockheed SR-71 Blackbird registration number 61-7972 took to fly from New York to London. The record was set in September 1974 as the high speed reconnaissance aircraft made its way to the Farnborough Airshow, its first public appearance outside America. (For comparison, the fastest time set by Concorde for that route was two hours and fifty-three minutes.) The twin-engined Blackbird was designed to fly fast enough and high enough to evade the defences of any country it was flying over. It cruised at speeds above three times the speed of sound and its normal operating height was 80,000ft, although it was capable of going higher. Unsurprisingly it set numerous records. It is officially the fastest manned jet aircraft achieving 2,193mph in 1976.

On its return flight from Farnborough 61-7972 flew London to Los Angeles in three hours, forty-seven minutes, thirty-nine seconds at an average speed of over 1,400mph. Its average speed was lower due to it having to slow down to refuel mid-air and being unable to fly supersonically over Great Britain.

A Blackbird at the International Air Tattoo in 1985. Those who saw the machine were impressed by its outstanding rate of acceleration. *Derek N. Ferguson*

In 1990, as the type began to be withdrawn from service the same aircraft was being delivered to an air museum in Washington and, in a show of blistering performance unlikely to be seen again, flew the 2,300 miles from Los Angeles in sixty-four minutes. By 1999 one of the most impressive of all aircraft was finally grounded.

✈ AIRCRAFT NICKNAMES

Aardvark	–	General Dynamics F-111
Armlong Woolworth Allsoggy	–	Armstrong Whitworth Argosy
Bognor Bloater	–	White & Thompson NT.3
Bone	–	Rockwell B-1
Buff (Big Ugly Fat Fellow)	–	Boeing B-52
Clunk	–	Douglas SBD Dauntless
Coffee Pot	–	Breguet IV
Connie	–	Lockheed Constellation
Death Angel	–	Boeing B-9
Double Ugly	–	McDonnell Douglas F-4 Phantom
Dragon Lady	–	Lockheed U-2
Dumbo	–	Supermarine S.24/37
Elastoplast Bomber	–	Messerschmitt Me 323 Gigant
Electric Jet	–	General Dynamics F-16
Elephant	–	Martinsyde G.100
Faithful Annie	–	Avro Anson
Flying Bedstead	–	Rolls-Royce Thrust Measuring Rig
Flying Porcupine	–	Short Sunderland
Flying Prostitute	–	Martin B-26 Marauder
Flying Suitcase	–	Handley Page Hampden
Gooneybird	–	Douglas C-47
Grandpappy	–	Boeing XC-105
Gunbus	–	Vickers FB.5
Habu	–	Lockheed SR-71
Harry Tate	–	Royal Aircraft Factory Re.8
Huey	–	Bell UH-1 Iroquois
Jug	–	Republic P-47 Thunderbolt
Meatbox	–	Gloster Meteor
Puff the Magic Dragon	–	Douglas AC-47
Pulpit	–	Royal Aircraft Factory BE.9
Son of a Bitch, 2nd Class	–	Curtis SB2C Helldiver
Squirt	–	Saunders-Roe SR/A1
Stability Jane	–	Royal Aircraft Factory BE.2c

Stringbag	–	Fairey Swordfish
Tam's Tram	–	de Havilland Heron
The Mortician's Friend	–	Lavochkin LaGG-3
Thud	–	Republic F-105 Thunderchief
Thunderscreech	–	Republic XF-84H
Tin Donkey	–	Junkers J-1
Tin Triangle	–	Avro Vulcan
Ugly Bus	–	Sikorsky S-92
Whispering Death	–	Bristol Beaufighter
Whispering Giant	–	Bristol Britannia
Whistling Wheelbarrow	–	Armstrong Whitworth Argosy
Widowmaker	–	Lockheed F-104 Starfighter
Wimpy	–	Vickers Wellington

NOSE ART

Since the First World War, combat aircraft have had paintwork applied to their forward fuselages. The Second World War saw a huge burst of creativity with many fighters and bombers having names assigned, along with figurative artwork, often of scantily clad women.

Royal Air Force

Halifax: Chesty, Clueless, Friday the 13th, Honey Chile, London's Revenge, Pete the Penguin, Pinocchio, Ruhr Valley Express, Willie the Wolf from the West, Vicky the Vicious Virgin.

Lancaster: Admiral Prune II, Dante's Daughter, Edith, Edna, Fannin' Fanny, Frederick III, Mickey the Moocher, Oor Wullie, Phantom of the Ruhr, Press on Regardless, Queen of Spades, Victorious Virgin, Whoa Bessie!, Zombie.

Stirling: Blondie, It's in the bag Jane, Just Jane, Per Lager Ad Aspro, Semper in Excreta, The Gremlin Teaser, Yorkshire II.

United States

A-20 Havoc: Battlin' Bob, Florida Gator, Good Time Charlie, Pistol Packin' Mama, Rough Stuff, Sweet Li'l Kitten, Yodel Bloomer.

B-17 Flying Fortress: Bad Penny, Bam Bam, Bomber Dear, Dog Breath, Doris Jr, Duckie, Earthquake McGoon, Fearless Fosdick, Hell's Cargo, Iza Vailable,

Knock-out Dropper, Little Tush, Memphis Belle, Pistol Packin' Mama, Shoo Shoo Shoo Baby, Snap! Crackle! Pop!, Thru Hel'en Hiwater, Thunder Bird, Two Beauts, Yankee Doodle Dandy, You've Had It Zombie.

B-24 Liberator: Axis Grinder, Bomb Boogie, Booby Trap, Dinah Might!, Dirty Gertie, Dragon Lady, Easy Maid, Frisco Frisky, Fuzzy Wuzzy, Indian Thummer, Joisey Bounce, Lady Diana, Lady Luck, Monotonous Maggie, My Heart Belongs to Daddy, Pleasant Surprise, Short Snorter, Strawberry Bitch, The Dragon and his Tail, The Wild Hair, We'll Get by, Wonder Gal.

B-25 Mitchell: Blonde Bomber, Calamity Jane, Grass Cutter, Incendiary Blonde, Milk Run, Seven Day Leave, The Gay Mare, The Ink Squirts, Wolf Bait.

B-29: Bock's Car, Double Exposure, Enola Gay, Impatient Virgin, Lucky Lady, Peace on Earth, Slick Chick's, Supine Sue, Tail Wind, The Cultured Vulture.

P-38 Lightning: Bambi, Big Nick, Da Quake, Fubar, Miss Mona, Princess Pat, Secret Weapon, Sleepy-Time Gal, Spirit of Oak Ridge, Stinger.

P-47 Thunderbolt: Arizona Pete, Big Ass Bird II, Bundle of Joy, California or Bust, Cripes A'Mighty!, Hairless Joe, Indianapolis, Lollapoluza, Miss Mary Lou, Miss Plainfield, The Deacon.

The B-17 'Shoo Shoo Shoo Baby' is on display at the National Museum of the United States Air Force. It flew on twenty-four combat missions during the Second World War. *Henrietta B. Ferguson*

P-51 Mustang: Bow Legs, Captain Marvel, Diablo, Ex-Lax ... Shht 'n' Git!, Little Skunk, Peanuts, Rusty, Sack Queen, Shack Lassie, Shangri-La, Stinky 2, Sweet Bet, Sweetie Face, Turnip Termite, Wheezy.

 BIG MIG

Around 13,500 Mikoyan-Gurevich MiG-21s were built. The Soviet Union-designed machine is the most widely operated jet fighter aircraft, having been used in over fifty countries:

Afghanistan
Algeria
Angola
Azerbaijan
Bangladesh
Belarus
Bulgaria
Burkina Faso
Cambodia
Chad
China
Congo
Croatia
Cuba
Czechoslovakia
East Germany
Egypt
Eritrea
Ethiopia
Finland
Georgia
Guinea
Guinea-Bissau
Hungary
India
Indonesia
Iran
Iraq
Israel

Laos
Libya
Madagascar
Mali
Mongolia
Mozambique
Namibia
North Korea
Nigeria
Poland
Romania
Serbia
Slovakia
Somalia
Sudan
Syria
Tanzania
Turkmenistan
Uganda
USA
USSR
Ukraine
Vietnam
Yemen
Yugoslavia
Zaire
Zambia
Zimbabwe

This MiG-21 served with the Czechoslovakian Air Force. It is a two-seat version, used for pilot training. *Derek N. Ferguson*

✈ THE MOST LUXURIOUS AIRCRAFT?

Private jets are the ultimate in status symbols. The world's wealthy are perfectly at home in the air when flying in Learjets, Gulfstreams or even larger craft. If money is no object there are few limits to the levels of luxury available. The ultimate private mode of aerial transport has to be the VIP Airbus A380, ordered by a client from the Middle East. Full details have not been disclosed but the converted airliner – rumoured to cost half a billion dollars – is reported to contain the following:

A private lift, able to be lowered to the airport tarmac,
Garage,
Stable,
Concert hall,
Steam room,
Boardroom,
Twenty first-class staterooms,
Four private suites,
'Wellbeing' room, floor of which displays real-time images of the terrain below,
Prayer room with virtual prayer mats to face Mecca.

✈ PINGERS, BAGGIES AND JUNGLIES

The Royal Navy's Fleet Air Arm is equipped with different types of helicopters able to perform specific roles. The crews and aircraft have been given nicknames: those engaged in anti-submarine warfare are known as 'pingers', airborne early warning are 'baggies' and troop carriers are called 'junglies'. The reasons behind these names? Sonar uses 'pings' to detect submarines under water. The airborne early warning helicopters use a rotating radar dish that hangs underneath the fuselage during flight, resembling a bag. Navy helicopters earned the nickname 'junglies' for operating in difficult conditions in Borneo in the 1960s.

✈ LUCKY ESCAPES: THE GALUNGGUNG GLIDING CLUB

In 1982 a British Airways flight from Kuala Lumpur to Australia was proceeding without incident when a choking black smoke started to appear within the cabin. Things got even worse when all four engines stopped. The Boeing 747 was at 37,000ft. Its crew had no option but to try to glide the aeroplane towards Jakarta, while attempting to restart the engines. If this was unsuccessful they faced a night-time ditching in the Indian Ocean. The captain, Eric Moody, made an announcement to the passengers: 'All four engines have stopped. We are all doing our damnedest to get them going again. I trust you are not in too much distress.' In the quiet cabin, many passengers wrote farewell notes to their loved ones. The 747 descended to a height of 12,000ft above the water before the Rolls-Royce engines came back to life. A relieved crew continued for Jakarta but their problems weren't over as they discovered they couldn't see the runway clearly. Whatever they had flown through had scratched the cockpit windows making them opaque. Moody was later asked what it was like to land in this way with restricted vision. He replied: 'It was a bit like negotiating one's way up a badger's arse.' It was found later that ash from the Mount Galunggung volcano had drifted into the flight path unseen. It swamped the engines that were only able to restart when the ash had solidified and fallen free. A group called the Galunggung Gliding Club was formed by crew and passengers.

✈ OPERATION CODENAME

Operation Black Buck: Long-range bombing/anti-radar missile raids, Falkland Islands, 1982.

Operation Bowler: Precision bombing raid on Venice, 1945.

Operation Carthage: Raid on Gestapo HQ, Copenhagen, 1945.

Operation Catechism: Last bombing operation to sink the *Tirpitz*, 1944.

Operation Centerboard I: The first atomic bombing mission against Japan.

Operation Chastise: The Ruhr Dams raid, 1943.

Operation Deny Flight: No-fly zone enforcement, Bosnia and Herzegovina, 1993.

Operation El Dorado Canyon: American bombing of Libya, 1986.

Operation Frantic: Ultra-long-range bombing missions from UK, Italy and Soviet Union, 1944.

Operation Gomorrah: The bombing of Hamburg, 1943.

Operation Jericho: Amiens Prison raid, 1944.

Operation Knicker: Initial airlift of supplies into Berlin, 1948.

Operation Taxable: Deception flights before D-Day by 617 Squadron, June 1944.

Operation Titanic: Deception missions involving the dropping of fake parachutists and pyrotechnics on D-Day, June 1944.

Operation Varsity: The dropping of airborne troops to secure the crossing of the Rhine, 1945.

✈ FORMATION FLYERS

Country	Team Name	Aircraft Flown
Canada	Snowbirds	Tutor
China	August 1st	Chengdu J-10
Egypt	Silver Stars	K-8E Karakorum
Finland	Midnight Hawks	BAE Systems Hawk
France	Patrouille Acrobatique de France (French Acrobatic Patrol)	Alpha Jet
Italy	Frecce Tricolori (Tricolour Arrows)	Aermacchi MB-339
Malaysia	Smokey Bandits	MiG-29
Pakistan	Sherdils	K-8 Karakorum
Poland	White-Red Sparks	TS-11 Iskra
Russia	Russian Knights	Sukhoi Su-27

In 2011 the Red Arrows flew an eight-ship formation due to the loss of Flt Lt Jon Egging. He was killed in a flying accident following a display at Bournemouth.
Norman H. Ferguson

	Russian Falcons	Su-27 and MiG-29
	Swifts	MiG-29
Saudi Arabia	Saudi Hawks	BAE Systems Hawk
Singapore	Black Knights	F-16
South Korea	Black Eagles	KAI T-50 Golden Eagle
UK	Red Arrows	BAE Systems Hawk
Ukraine	Ukrainian Falcons	MiG-29
USA	Blue Angels	F/A-18
	Thunderbirds	F-16
Japan	Blue Impulse	Kawasaki T-4
Spain	Patrulla Águila (Eagle Patrol)	Casa C-101 Aviojet
Switzerland	Patrouille Suisse (Swiss Patrol)	Northrop F-5E Tiger II
Turkey	Turkish Stars	Northrop NF-5
United Arab Emirates	Al Fursan (The Knights)	Aermacchi MB-339NAT

✈ ABBREVIATIONS AND ACRONYMS

AGL	–	Above Ground Level
AEW	–	Airborne Early Warning
AWACS	–	Airborne Warning and Control System
BVR	–	Beyond Visual Range
CAP	–	Combat Air Patrol
CFIT	–	Controlled Flight Into Terrain
G-LOC	–	G-induced Loss of Consciousness
EWO	–	Electronic Warfare Officer
FFAR	–	Folding Fin Aerial Rockets
FIDO	–	Fog Investigation Dispersal Operation
HOTAS	–	Hands On Throttle-and-Stick
HUD	–	Heads Up Display
IAS	–	Indicated Air Speed
ILS	–	Instrument Landing System
INS	–	Inertial Navigation System
IP	–	Initial Point
J-STARS	–	Joint Surveillance Target Attack Radar System
LGB	–	Laser Guided Bomb
MFD	–	Multi-Function Display
PGM	–	Precision-Guided Munition
QRA	–	Quick Reaction Alert
RIO	–	Radar Intercept Officer
SAM	–	Surface-to-Air Missile
SARBE	–	Search and Rescue Beacon Equipment
TACAN	–	Tactical Air Navigation System
TFR	–	Terrain Following Radar
UAV	–	Unmanned Aerial Vehicle
VOR	–	VHF Omni-directional Radio
V/STOL	–	Vertical and/or Short Take-off and Landing

✈ THIS IS YOUR 'CADETTE' SPEAKING

Potential names for what we now call 'flight attendants':

agent	airmaid
airaide	aidette
airess	attaché
airette	attendant

cadette
courier
courierette
escort

hostess
page
purserette
skipper

# THE RAF IN 1939

UK-based commands: Bomber Command; Fighter Command; Coastal Command; Training Command; Maintenance Command; Reserve Command; Balloon Command.

# CONCORDE LOCATIONS

After their retirement in 2003, British Airways' fleet of seven Concordes were distributed to various locations.

Registration	Location	Country	Notes
G-BOAA	Museum of Flight, East Fortune	United Kingdom	Aircraft restored inside hangar.
G-BOAB	Heathrow Airport	United Kingdom	Kept at British Airways engineering facility.
G-BOAC	Manchester Airport	United Kingdom	Aircraft is kept in a specially built hangar.
G-BOAD	Intrepid Sea, Air and Space Museum, New York	USA	This Concorde is kept in the open air beside the Hudson River.
G-BOAE	Grantley Adams Airport	Barbados	G-BOAE is preserved in a specially built hangar.
G-BOAF	Filton	United Kingdom	Aircraft currently not open to visitors.
G-BOAG	Museum of Flight, Seattle	USA	Aircraft kept outside at the museum.

 WORDS OF WISDOM

'I think it is well also for the man in the street to realise that there is no power on earth that can protect him from being bombed, whatever people may tell him. The bomber will always get through.'

Stanley Baldwin MP, November 1932.

 CON AIR

Fictional airlines from the big screen:

Columbus Airlines	–	*Airport 1975*
Federation World Airlines	–	*The Concorde... Airport '79*
Fuji Air	–	*Die Hard 2*
Intercity Air	–	*Fearless*
Mahigan Air	–	*The Grey*
North East Airlines	–	*Die Hard 2*
Oceanic Airlines	–	*Executive Decision*
Plymouth Air	–	*Knowing*
Royce Air	–	*Panic in the Skies*
South Pacific Airlines	–	*Snakes on a Plane*
Trans American	–	*Airplane*
Trans Global Airlines	–	*Airport*
Volée Airlines	–	*Final Destination*
Windsor Air	–	*Die Hard 2*

TO RUSSIA WITH LIFT

Aircraft landing outside airfields is unusual and newsworthy. When one lands in the middle of Moscow during the Cold War, it's a world-wide sensation. In 1987 a young German man called Mathias Rust set off from Helsinki in a Cessna 172 light aircraft, registration D-ECJB. The 19-year-old's course took him east through the Iron Curtain and into Warsaw Pact territory. He was spotted and Soviet jets were sent up to investigate but his journey was uninterrupted. Astonished onlookers watched as the small white plane circled over the Russian capital before settling in to land in Red Square. To say the Soviet air defence commanders were embarrassed is an understatement. It was later claimed because it was a public holiday the air defenders' guards were down. Rust was imprisoned but later released,

The Cessna 172 flown by Mathias Rust is now in the Deutsches Technikmuseum, Berlin. It spent time displayed outside a sports club in Japan before being returned to Germany. *Andrey Belenko*

and he said afterwards that his flight, which was intended to further world peace, was 'irresponsible'. He never piloted an aircraft again.

✈ LUCKY ESCAPES: LIGHTNING BOLT

Wing Commander 'Taffy' Holden was an engineering officer on a maintenance unit at RAF Lyneham. He was also a licensed pilot, able to fly single-engined propeller aircraft. On 22 July 1966 no qualified Lightning pilot could be found to carry out ground tests on a malfunctioning machine, serial number XM 135, so Wing Commander Holden climbed into the single-seat cockpit. Everything proceeded to plan until on the third run when he inadvertently selected reheat. The Lightning was an aircraft with rapid acceleration and before he knew much about it, Holden found himself in the air piloting an aircraft he was not trained to fly. The simple solution to his predicament would be to eject using the rocket-powered seat. However as no one expected the seat to be needed the safety pins were still in place. Holden later said he faced death five times during his flight. He had to abandon landing approaches when he started losing height, but eventually became more confident in handling the 11-ton fighter jet and was able to land safely. Although Holden survived the unexpected twelve-minute flight without physical injury he later required treatment for its psychological effects.

 # TWO WINGS GOOD, FOUR WHEELS BAD

Aircraft usually, but not always, come out on top when pitted against cars:

Participant (aircraft)	Participant (car)	Race details	Location	Year	Winner
RAF Eurofighter Typhoon	Bugatti Veyron	Distance: 2 miles. Car goes down runway for 1 mile, turns and returns to the starting line. Aircraft takes off, pulls up to 1 mile in altitude and returns.	RAF Coningsby, for BBC *Top Gear*	2008	Aircraft
F/A-18A Hornet, from Blue Angels team	Chevrolet Corvette ZR1 'Blue Devil'	Distance: 1 mile	Pensacola Naval Air Station, Florida	2009	Aircraft
Tornado	Lamborghini Reventon	3,000m runway	Ghedi military airfield, Italy	2007	Aircraft
F-16	Spyker F8-VII Formula 1 car	Distance: 1km	Volkel air force base, Netherlands	2007	Aircraft
Harrier	Audi R10 Le Mans car	Distance: 1km	RAF Wittering, as part of the Goodwood Festival of Speed	2006	Aircraft

CF-18 Hornet	Ferrari Enzo	Runway	Cold Lake airbase, Canada	2010	Aircraft
Italian Air Force Eurofighter Typhoon	Ferrari F2003-GA Formula 1 car	Car driven by Michael Schumacher. Race consisted of three straight-line speed contests: 600m, 900m, 1,200m.	Aeronautica Militare air base, Grosseto, Italy	2003	Car, aircraft, aircraft
MiG-29	Williams Formula 2 car	Distance: 600m	Krumovo AFB/Plovdiv Airport	2011	Car

 BANNED IN THE EU

Airlines banned from operating in European airspace, as at April 2012:

Afghanistan: Ariana Afghan Airlines, Kam Air, Pamir Airlines, Safi Airways.

Angola: Aerojet, Air26, Air Gicango, Air Jet, Air Nave, Angola Air Services, Diexim, Fly 540, Gira Globo, Heliang, Helimalongo, Mavewa, Sonair.

Benin: Aero Benin, Africa Airways, Alafia Jet, Benin Golf Air, Benin Littoral Airways, Cotair, Royal Air, Trans Air Benin.

Congo: Aero Service, Equaflight Services, Societe Nouvelle Air Congo, Trans Air Congo, Equatorial Congo Airlines SA.

Democratic Republic of Congo: African Air Services Commuter, Air Kasai, Air Katanga, Air Tropiques, Blue Airlines, Bravo Air Congo, Business Aviation, Busy Bee Congo, Cetraca Aviation Service, CHC Stellavia, Congo Express, Compagnie Africaine d'Aviation, Doren Air Congo, Enterprise World Airways, Filair, Galaxy Kavatsi, Gilembe Air Soutenance, Goma Express, Gomair, Hewa Bora Airways, International Trans Air Business, Jet Congo Airways, Kin Avia, Korongo Airlines, Lignes Aeriennes Congolaises, Malu Aviation, Mango Aviation, Safe Air Company, Services Air, Stellar Airways, Swala Aviation, TMK Air Commuter, Tracep Congo Aviation, Trans Air Cargo Services, Wimbi Dira Airways, Zaabu International.

Djibouti: Daallo Airlines.

Equatorial Guinea: Cronos Airlines, Ceiba Intercontinental, Punto Azul.

Gabon: Afric Aviation, Air Services SA, Air Tourist (Allegiance), Nationale et Regionale Transport (Nationale), SCD Aviation, Sky Gabon, Solenta Aviation Gabon.

Indonesia: Air Pacific Utama, Alfa Trans Dirgantata, Asco Nusa Air, Asi Pudjiastuti, Aviastar Mandiri, Dabi Air Nusantara, Deraya Air Taxi, Derazona Air Service, Dirgantara Air Service, Eastindo, Enggang Air Service, Ersa Eastern Aviation, Gatari Air Service, Indonesia Air Transport, Intan Angkasa Air Service, Johnlin Air Transport, Kal Star, Kartika Airlines, Kura-Kura Aviation, Lion Mentari Airlines, Manunggal Air Service, Merpati Nusantara Airlines, Mimika Air, National Utility Helicopter, Nusantara Air Charter, Nusantara Buana Air, Nyaman Air, Pelita Air Service, Penerbangan Angkasa Semesta, Pura Wisata Baruna, Riau Airlines, Sampoerna Air Nusantara, Sayap Garuda Indah, Sky Aviation, SMAC, Sriwijaya Air, Survei Udara Penas, Surya Air, Transnusa Aviation Mandiri, Transwisata Prima Aviation, Travel Express Aviation Service, Travira Utama, Tri-MG Intra Asia Airlines, Trigana Air Service, Unindo, Wing Abadi Airlines.

Kazakhstan: Aero Air Company, Air Almaty, Air Trust Aircompany, AK Sunkar Aircompany, Asia Continental Airlines, Asia Wings, ATMA Airlines, Avia-Jaynar/Avia-Zhaynar, Beybars Aircompany, Berkut Air/Bek Air, Burundayavia Airlines, Comlux, Deta Air, East Wing, Eastern Express, Euro-Asia Air, Euro-Asia Air International, Fly Jet KZ, Investavia, Irtysh Air, Jet Airlines, Jet One, Kazair Jet, Kazairtrans Airline, Kazairwest, Kazaviaspas, Mega Airlines, MIRAS, Prime Aviation, Samal Air, Sayakhat Airlines, Semeyavia, SCAT Airlines, Skybus, Skyjet, Ust-Kamenogorsk/Air Division of EKA, Zhetysu Air Company.

Kyrgyzstan: Air Manas, Asian Air, Avia Traffic Company, Aerostan (ex-Bistair-Fez Bishkek), Central Asian Aviation Services (CAAS), Click Airways, Dames, Eastok Avia, Itek Air, Kyrgyz Trans Avia, Kyrgyzstan, Kyrgyzstan Airline, S Group Aviation, Sky Way Air, Trast Aero, Valor Air.

Mauritania: Mauritania Airlines.

Mozambique: Mozambique Airlines, Mozambique Express/Mex, Trans Airways/Kaya Airlines, Helicopteros Capital, CFA Mozambique, Unique Air Charter, Aerovisao de Mozambique, Safari Air, ETA Air Charter LDA, Emilio Air Charter LDA, CFM-TTA SA, Aero-Servicos SARL, VR Cropsprayers LDA.

Philippines: Aeroequipment Aviation, Aeromajestic, Aerowurks Aerial Spraying Services, Air Asia Philippines, Air Philippines Corporation, Air Wolf Aviation Inc., Airtrack Agricultural Corporation, Asia Aircraft Overseas Philippines Inc., Aviation Technology Innovators Inc., Aviatour's Fly'n, Ayala Aviation Corp, Beacon, Bendice Transport Management Inc., Canadian Helicopters Philippines Inc., Cebu Pacific Air, Certeza Infosystems Corp., Chemtrad Aviation Corporation, CM Aero Services, Corporate Air, Cyclone Airways, Far East Aviation Services, F.F. Cruz and Company Inc., Huma Corporation, Inaec Aviation Corp, Interisland, Island Aviation, Island Transvoyager, Lion

Air Incorporated, Macro Asia Air Taxi Services, Mid-Sea Express, Mindanao Rainbow Agricultural Development Services, Misibis Aviation & Development Corp, Omni Aviation Corp, Pacific East Asia Cargo Airlines Inc., Pacific Airways Corporation, Pacific Alliance Corporation, Philippine Airlines, Philippine Agricultural Aviation Corp, Royal Air Charter Services Inc., Royal Star Aviation Inc., South East Asia Airline Inc. (SEAIR), Southern Air Flight Services, Southstar Aviation Company, Spirit of Manila Airlines Corporation, Subic International Air Charter, Subic Seaplane Inc., Topflite Airways, Transglobal Airways Corporation, World Aviation Corp, WCC Aviation Company, Yokota Aviation Inc., Zenith Air Inc., Zest Airways Incorporated.

Rwanda: Silverback Cargo Freighters.

Sao Tome and Principe: Africa Connection, British Gulf International Company Ltd, Executive Jet Services, Global Aviation Operation, Goliaf Air, Island Oil Exploration, STP Airways, Transafrik International Ltd, Transcarg, Transliz Aviation (TMS).

Sierra Leone: Air Rum Ltd, Destiny Air Services Ltd, Heavylift Cargo, Orange Air Sierra Leone Ltd, Paramount Airlines Ltd, Seven Four Eight Air Services Ltd, Teebah Airways.

Sudan: Alfa Airlines, Almajal Aviation Service, Almajara Aviation, Attico Airlines (Trans Attico), Azza Transport Company, Bader Airlines, Forty Eight Aviation, Green Flag Aviation, Marsland Company, Nova Airlines, Sudan Airways, Sudanese States Aviation Company, Sun Air Company, Tarco Airlines.

Suriname: Blue Wing Airlines.

Swaziland: Swaziland Airlink.

Venezuela: Conviasa.

Zambia: Zambezi Airlines.

✈ THE BOLD AVIATOR LAY DYING

(To the tune: 'My Bonnie Lies Over the Ocean')

'Oh, the bold aviator was dying
And as 'neath the wreckage he lay, he lay
To the sobbing mechanics about him
These last parting words he did say:

'"Take the cylinders out of my kidneys
The connecting rod out of my brain, my brain
From the small of my back get the crankshaft
And assemble the engine again.

'"Two valve springs you'll find in my stomach,
Three spark plugs are safe in my lung, my lung,
The prop is in splinters inside me,
To my fingers the joy-stick has clung.

'"Take the propeller boss out of my liver,
Take the aileron out of my thigh, my thigh,
From the seat of my pants take the piston,
Then see if the old crate will fly."'

 WORDS OF WISDOM

'There is still a newness about air travel, and, though statistics demonstrate its safety, the psychological effect of having a girl on board is enormous.'

US airline magazine, 1935, commenting on the advent of stewardesses.

SURVIVAL CLUBS

Caterpillar Club: Members can be any aircrew who have had their lives saved by using a parachute to escape a stricken aircraft.

Lieutenant J.T. 'Terry' Kryway uses a Martin Baker seat to eject from his Crusader fighter jet which caught fire while landing on the USS *Franklin D. Roosevelt*. *L.J. Cera, US Navy*

Goldfish Club: Those who have survived a ditching or parachuting into water.

Late Arrivals Club: Allied airmen who were able to reach their own lines after being forced down over enemy-held territory.

Ejection Tie Club: Those who have used a Martin Baker ejection seat to 'bang out' of an aircraft.

 # THE THIN BLUE LINE

Fighter Command Order of Battle – 1 August 1940:

Hurricane Squadrons
10 Group: 87, 213, 238
11 Group: 1, 17, 32, 43, 56, 85, 111, 145, 151, 257, 501, 601, 615
12 Group: 46, 229, 242
13 Group: 3, 73, 79, 232, 245, 249, 253, 263, 504, 605, 607

Spitfire Squadrons
10 Group: 92, 152, 234, 609
11 Group: 41, 64, 65, 74, 266, 610
12 Group: 19, 66, 222, 611
13 Group: 54, 72, 602, 603, 616

Blenheim Squadrons
10 Group: 604
11 Group: 25, 600, 604, Fighter Interception Unit
12 Group: 23, 29
13 Group: 219

Defiant Squadrons
12 Group: 264
13 Group: 141

Gladiator Squadron
13 Group: 804 Fleet Air Arm

Fulmar Squadron
13 Group: 808 Fleet Air Arm

10 Group covered the South West of England and South Wales area with airfields at Middle Wallop and Filton.

11 Group covered the South East of England with airfields at Biggin Hill, Kenley, Hornchurch, Debden, North Weald, Northolt, and Tangmere.

12 Group covered the Midlands and North Wales with airfields at Duxford, Coltishall, Kirton-in-Lindsey, Digby, and Wittering.

13 Group covered northern England, Northern Ireland and Scotland with airfields at Church Fenton, Catterick, Usworth, Turnhouse, Dyce, Wick, Prestwick, and Aldergrove.

DID YOU KNOW?

535:
Number of aircraft that attended the 2003 International Air Tattoo airshow at RAF Fairford, leading to it being officially recognised as the world's largest military airshow by the Guinness World Records.

SELF-HELP BOOKS

Fear of Flying by J. Allen Scott

Flying without Fear: 101 Fear of Flying Questions Answered by Paul Tizzard and Richard Conway

Conquering Your Fear of Flying by Maeve Byrne-Crangle

Conquer Your Fear of Flying: Post 9/11 Edition by Maeve Byrne-Crangle

Flying Without Fear by Captain Keith Godfrey

Flying Without Fear by Ivo Velli

Flying Without Fear by Duane Brown

Flying Fear Free: 7 Steps to Relieving Air Travel Anxiety by Sandra M. Pollino

Overcome Your Fear of Flying by Professor Robert Bor, Dr Carina Eriksen and Margaret Oakes

The Easy Way to Enjoy Flying by Allen Carr

Take the Fear Out of Flying by Maurice Yaffe

White Knuckles: Getting over the Fear of Flying by Layne Ridley

Flying, No Fear!: Conquer Your Fear of Flying by Adrian Akers-Douglas

How to beat your fear of flying: Advice, information and strategies that work by Marshall Collings

Fly Away Fear: Overcoming Your Fear of Flying by Elaine Iljon Foreman and Lucas van Gerwen

✈ FAMOUS FEATS OF AVIATION: OPERATION BLACK BUCK

In early 1982 Britain found itself in conflict with Argentina over the Falkland Islands. As a naval task force sailed south, plans were made to launch long-range bombing raids on the main airfield at Port Stanley. On 30 April the first of these missions was launched. The bomber was an Avro Vulcan, one of Britain's former nuclear V-bombers, that had first flown in the early 1950s.

A complicated series of refuellings took place with eleven Victor tankers refuelling each other in order to provide a single aircraft to carry on with the Vulcan. Eventually the bomber was alone. To avoid radar detection it descended to 300ft above the sea 40 miles out from its target. It then climbed to the bombing altitude of 10,000ft. The element of surprise was achieved and the Vulcan dropped its load of 21 x 1,000-pound bombs. The release point was 3 miles from the airfield target.

On the way north, the Vulcan was picked up by the air-defence radars of the British Task Force. Before any friendly-fire incident could take place the radio operator transmitted the code word for a successful mission: 'Superfuse'. Vulcan XM607 landed back at Ascension Island, completing the longest ever bombing raid at that time, lasting almost sixteen hours and covering 8,000 miles. Photo reconnaissance photographs showed that the first bomb of the twenty-one dropped had landed on the runway.

✈ VIRGIN NAMES

Each Virgin Airways airliner carries a painted representation of a woman, in the style of Second World War nose art, along with a name for each aircraft:

Cosmic Girl – Dancing Queen – Diana – English Rose – Hot Lips – Lady Luck – Lady Penelope – Ladybird – Mademoiselle Rouge – Maiden Tokyo – Miss Behavin – Miss Kitty – Mustang Sally – Mystic Maiden – Pretty Woman – Queen of the Skies – Ruby Tuesday – Silver Lady – Soul Sister – Spirit of Sir Freddie – Surfer Girl – Tinker Belle – Tubular Belle – Varga Girl – Virginia Plain

✈ GREAT PLANES: X-15

The North American X-15 is the fastest piloted aircraft ever built. It could reach heights that earned its pilots astronaut wings. The rocket-propelled craft was carried aloft by a B-52 carrier aircraft to around 45,000ft when it was then released. Safely clear, the pilot then ignited the single XLR-99 engine and unleashed 57,000 pounds of thrust. The rocket engine had only enough fuel for a two-minute burn but phenomenal performance figures could still be reached. In 1963 an X-15 reached an altitude of 354,200ft (almost 67 miles). In 1967 it went as fast as Mach 6.7 (4,520mph). Both of these became unofficial world records. Flying to the edges of the atmosphere and beyond, the pilots were able to see the curvature of the Earth and a sky that had changed into a deeper shade of blue. One pilot reported that while flying over the Las Vegas area he was able to see as far as Puget Sound in Washington state, a thousand miles away.

After the rocket burn, with the fuel depleted, the pilot would glide the machine back to a landing on the desert. As with all flying of this kind, it was not without danger and Major Michael Adams was killed when his X-15 lost control at high altitude and broke up in mid-air.

Neil Armstrong stands beside an X-15 following a test flight. Armstrong made seven flights in the hypersonic aircraft, including one where he made a successful landing despite the X-15 racing past the landing area at high altitude and heading towards Los Angeles. *NASA*

✈ FAMOUS FEATS OF AVIATION: 'WRONG WAY' CORRIGAN

Douglas Corrigan was an American pilot who made his name flying the Atlantic. His fame stemmed from supposedly doing this in error. As an engineer, he had worked on the construction of Charles Lindbergh's aircraft *Spirit of St Louis*. Following Lindbergh's success Corrigan resolved to make his own transatlantic flight and to fly to his ancestral home of Ireland, but his attempts to obtain official permission were denied. On 17 July 1938 he took off from Brooklyn and headed east. Twenty-eight hours later he landed in Dublin. Corrigan claimed he had been following the wrong end of the compass needle. He became famous for his unlikely feat, receiving a ticker-tape parade in New York. The *New York Post*'s headline 'Hail Wrong Way Corrigan' was printed in reverse in tribute. It was heavily suspected by the authorities that the experienced pilot Corrigan had flown eastwards deliberately, but he never admitted his 'error'.

✈ 'PISS OFF BIGGLES'

In the late 1980s a Welsh farmer became annoyed at the amount of times his isolated hilltop farm was being visited by low-flying military aircraft. It seemed as if the pilots were using his farm buildings as a practice target. Infuriated, he painted 'Piss off Biggles' on the roof of one of his buildings. For a while it seemed to attract more overflights from curious flyers, but eventually the RAF gave him a wider berth.

✈ ANCIENT AND TIRED AIRMEN

The ATA – officially the Air Transport Auxiliary and not the nickname above – was formed in 1939 to perform flying duties for the air forces by pilots who were not required for military service. The civilian organisation was initially set up to carry mail or passengers, including wounded service personnel and VIPs, but soon started the role it would become known for: ferrying aircraft to front-line squadrons. Due to the shortage of qualified pilots it was decided to enrol women pilots, a move that did not meet with unreserved acceptance. The women's section was led by Pauline Gower and one of her pilots was the aviator Amy Johnson, who was unfortunately to become the first ATA fatality. During the war ATA pilots performed 308,000 ferry flights, delivering 147 aircraft types.

✈ 633 SQUADRON

This 1964 Hollywood film is regarded as a classic aviation movie. It features all the standard ingredients of a war movie of the time: tough squadron commander, even tougher senior commander, dependable subordinate crewmen, love interest sub-plot with a beautiful woman, stirring theme tune, dramatic scenes in the air. Although Cliff Robertson's name was above the titles, the real star was the de Havilland Mosquito. Several flying examples were gathered for the aerial sequences and, in an action that seems sacrilegious now, one was deliberately destroyed for the film. It must be said, there are also some model shots that look as though an Airfix kit has been flung round by a fishing rod, but this doesn't detract from a determined film that reflects some of the dash and derring-do of the real Mosquito crews. The film's class is reflected in two moments. On the climatic operation the squadron are decimated in achieving their objectives. The tough squadron commander, played by Cliff Robertson, crash-lands his plane but the hero of the moment is not him but his navigator and a Norwegian onlooker who help pull him out of the wreckage. The second moment is at the very end. The waiting senior commander hears news of the successful raid and the heavy casualties. A subordinate officer tells him that the whole squadron may have been sacrificed. The senior commander replies: 'You can't kill a squadron' and is driven off into the proverbial sunset as the orchestral music plays the film out.

✈ WHAT'S YOUR POISON?

These were the alcoholic drinks available to passengers flying with British Overseas Airways Corporation in the early 1970s:

Miniatures:
Sherry, Vermouth (sweet or dry) – 15p
Martini Cocktail (sweet or dry) – 30p
Scotch, American and Canadian Whiskies – 30p
Gin, Rum, Vodka, Cognac, Drambuie, Cointreau – 30p

Wines:
Champagne – 65p
French wine (red or white) – 30p

Beers:
English beers – 15p
Lager – 15p

AIRCRAFTMAN LAWRENCE OF ARABIA

Thomas Edward Lawrence is known for his daring exploits in the Middle East during the First World War. He took part in the Arab Revolt and famously took the port of Aqaba in a surprise attack. What is less well known is that T.E. Lawrence spent longer in the RAF than the army. Following his successes in the desert he became famous and the press enthusiastically followed his every move. Desiring a quieter life, he joined the RAF in 1922 as 'John Hume Ross'. His cover was blown by the press and he left, whereupon he enlisted in the army's Tank Corps, this time after legally changing his name to T.E. Shaw. He petitioned to rejoin the air force and in 1925 was wearing the light blue uniform again. As part of his service Lawrence/Shaw worked as a mechanic, developing the RAF's air-sea rescue speedboats, following his witnessing of a fatal crash off Plymouth Sound. He also worked on the preparations for the 1929 Schneider Cup seaplane race. He enjoyed life in the air force and once wrote: 'The RAF's solidity and routine have been anchors holding me to life and the world.' It wasn't to last and in 1935 his period of service was up and he returned to civilian life. Three months later Lawrence was riding one of his beloved motorcycles when he was involved in an accident. He went into a coma and never recovered, dying on 19 May 1935, aged 46.

 SRU-21/P AVIATOR'S SURVIVAL VEST

Survival equipment for aviators in the early days consisted mainly of wearing a flat cap and goggles. Flightwear developed alongside machines until in the 1960s American aircrew flying in Vietnam wore a survival vest that included these items:

Smith and Wesson .38 calibre pistol with 6 rounds tracer ammunition and 17 rounds ball ammunition; knife; emergency radio; foliage-penetrating signal kit; signalling mirror; whistle; light sticks; compass; waterproof matches; butane lighter; torch; water storage bag; fishing net; tourniquet; blanket; insect repellent; snake bite kit; insect sting kit; camouflage kit; survival manual.

✈ FLUTTER TEST

Test flying is carried out on all aircraft that roll out from manufacturing plants to ensure they are suitable for handing over to the client, be they a civil operator or a military unit. The following test was performed on the BAE Systems 146 series of four-engined airliners to gauge the reactions to adverse and sudden forces on the aircraft's control surfaces and ensure no undamped oscillations were excited. In other words, to make sure if it hit a pocket of turbulent air it wouldn't go out of control. While scientific, the methods of simulating such an event might appear surprising:

Test Procedure:
Aircraft not to exceed 354 knots indicated air speed at or above 18,500ft.
Fuel contents should not be less than 12,000 pounds.
Flaps – set to zero.
Landing gear – set to up.
Test to begin at 19,200ft, speed 350 knots.
Hammer to be applied to control column (pilot's).
Hammer to be applied to control column (co-pilot's).
Rudder pedal to be stamped with foot.

ACES HIGH

Roland Garros was a pioneering French aviator, famous for making the first flight across the Mediterranean in 1913. In the First World War he became the first true fighter pilot. His Morane-Saulnier monoplane was fitted with a fixed machine gun, pointing forwards on top of the engine cowling. It was aimed at the target by the pilot manoeuvring the aircraft, rather than having an observer firing a movable gun. If unmodified, there was an immediate flaw with this set up: bullets would impact the propeller. To prevent this Garros' aircraft had an innovative system. Triangular metal plates were fitted to the propeller blades to deflect any bullets. While on patrol on 1 April 1915 Garros spotted a German biplane and opened fire from 12 yards. The Albatross' petrol tank caught fire and the aircraft soon crashed. Garros landed nearby but found the German crew dead.

Roland Garros went on to shoot down two more German aircraft until being forced to land behind enemy lines, where his device was discovered. He escaped from captivity in 1918 and took to the air again scoring more victories. Garros was shot down and killed a month before the end of the war, having shot down four aircraft. In 1928 a Parisian tennis club gifted land to build a stadium, a condition being that it was named after one of their ex-members. The Stade Roland Garros is now the home of the French Open.

✈ PRESENTATION SPITFIRES

During the Second World War the public were keen to help the war effort. Fund-raising for the building of aircraft was one way of doing this and towns, companies, individuals and others donated money. £5,000 bought a Spitfire and although other aircraft were provided for, the bulk were Supermarine's classic fighter. Enough money was raised in Kent to fund a whole squadron: 131 (County of Kent). One of the most unexpected donations came from a German pilot who had been recently shot down by RAF fighters, who put a five-mark note in a collecting box while being taken off to a prisoner-of-war camp. Listed are some of the groups who sent money for Spitfires:

Association of Men of Kent and
    Kentish Men
Blackpool
Borough of Islington, London
Breeders and owners of pedigree
    dogs
British Reinforced Concrete
Brooke Bond tea manufacturer
Bryant and May match makers
Captain Leslie Gamage
Cooperative Wholesale Society
Covent Garden Market
Danish residents in Britain
Dunlop tyre maker
Edwin Bacon, Grimsby trawler
    owner
General Electric Company
Gillingham
Great Western Railway
Greyhound Racing Association

Hendon
Hinckley and District Hosiery
    Manufacturers
Iron and Steel Trades Confederation
James Lathan, Timber Importers
Kennel Club
Leicester
London Stock Exchange
Lytham St Annes
Marks & Spencer
Newfoundland
Norfolk Farmers Union
Observer Corps
People named Dorothy
Stornoway
Sunderland
The Gold Coast
The Nizam of Hyderabad
Window cleaners of Britain
Woolworths

✈ LIGHTNING FLIGHT

The English Electric Lightning was the ultimate expression of British supersonic fighter jet design. Designed in the 1950s its role was that of interceptor, intended to climb rapidly to high altitudes to meet incoming Soviet bombers. However, the twin-engined fighter was notoriously short of endurance. A typical sortie would last forty minutes and a flight that saw liberal use of the engines' reheat could lower that to ten or twelve minutes. It was retired from RAF service in 1988 – a sad day for those who admired its pugnacious yet sleek appearance, its stunning 'rotation' take-offs and who also saw it as a link to the post-war glory days of British aviation. A few years before its retirement, an ex-RAF ground radar operator was given the opportunity to fly in a two-seat Lightning T5 from the last base operating the type – RAF Binbrook in Lincolnshire. He described what it was like to achieve a boyhood ambition in the last all-British supersonic fighter:

'As I walked out to the aircraft it seemed to get bigger and more menacing in its drab grey and green camouflage. The Lightning was an unusual design in that the engines were fitted vertically. As a result the cockpit looked very high off the ground. Jake, my pilot for the trip, did the walk-round checks as I clumsily climbed the steps, weighed down with all the clothing and kit required for fast-jet flight: flying suit, one-piece immersion suit, safety harness, leg restraints, G-suit and "bone dome" helmet to name just some. Once in I got myself settled into the ejection seat. Jake had said the day before: "If I say eject – don't wait for an answer." He would be gone, and was worth more to the air force than I was. A sobering thought.

'It took time to get my seat harness secured, leg restrainers affixed, personal equipment connector attached and the seat's safety pins stowed away. The small, dark cockpit got even smaller when Jake settled into his own ejector seat. His hands soon became a blur as they moved across the myriad of instruments and switches. With a final safety brief carried out both engines were started, one more reluctant to cooperate than the other. The large canopy was closed with a reassuring clunk. We were ready.

'After being given permission to move from the dispersal area we moved off. Brakes were immediately tested and clearance was given to line up on the main runway.

'Face mask clipped into place. Visor lowered. Last check of the safety harness. All systems checked and confirmed correct.

'Throttles were opened and we held on 90 per cent power for a few seconds before the brakes were released. The Lightning, realising it was free, leapt forward. The throttles were advanced to full 'cold' power for two seconds before reheat was engaged. I felt the force of acceleration pushing me back into the seat. The Rolls-Royce Avons were now producing around 33,000lb of thrust. In no time at all we were airborne. Undercarriage was selected up

Lightning T.5 XS458 at RAF Binbrook in April 1983. After retiring, XS458 passed into private ownership and is used for ground run displays. *Raymond A. Ferguson*

quickly to avoid it being overstressed in the rapidly increasing airflow. After gaining enough speed Jake pulled the control column back. Our nose went up to an angle nearing the vertical.

'We reached 30,000ft within a few minutes.

'After such an exhilarating take-off and climb we levelled out. I now had a chance to catch my breath and try and relax a little. I was asked if I would like to try my hand at flying. I carefully took hold of the stick. The aircraft was very responsive and gentle inputs were all that were needed. My five minutes were soon up however. The pilot wanted his jet back. I wasn't complaining as next up was the moment I had waited for. A long-held boyhood ambition: supersonic flight – breaking the Sound Barrier. Once almost mythical, it was commonplace for a machine like the Lightning.

'The nose was eased down slightly. Reheat selected. We started to accelerate. As the ground was covered in low cloud there was no real sensation of speed. The only indication was the Mach meter, which jumped as we reached Mach 1.0. A simple change on a small instrument marking a dream realised. We continued, gaining speed, getting faster and faster, reaching Mach 1.5 in a matter of seconds. I was now travelling over 1,000mph. However, the Lightning's short endurance is never far away from the pilot's mind. Reheat was deselected and the dive break extended to slow us down. Jake was planning our let-down procedures to take us earthwards back to base when he was advised that a Vulcan air-refuelling tanker was in the area and looking for trade. Good news for us.'

Flt Lt R.A. Ferguson, RAF VR(T)

 HIGH FLIGHT

'Oh! I have slipped the surly bonds of Earth
And danced the skies on laughter-silvered wings;
Sunward I've climbed, and joined the tumbling mirth
Of sun-split clouds – and done a hundred things
You have not dreamed of – wheeled and soared and swung
High in the sunlit silence. Hov'ring there,
I've chased the shouting wind along, and flung
My eager craft through footless halls of air ...

'Up, up the long, delirious, burning blue
I've topped the wind-swept heights with easy grace
Where never lark nor even eagle flew –
And, while with silent, lifting mind I've trod
The high, untrespassed sanctity of space,
Put out my hand, and touched the face of God.'

John Gillespie Magee Jr.

Pilot Officer Magee started composing this famous poem while in the air, flying a Spitfire Mark V on a high-altitude test flight. He was killed four months later in an air accident. He was 19.

✈ WORDS OF WISDOM

'I have flown everything from a 65hp Piper J-3 Cub to all the USAF Century Series fighters. But there is still something about being in the back seat of a yellow J-3 Cub with the door open on a glass smooth early morning just after sunrise with the slipstream hitting you in the face and Mother Earth slowly – I mean slowly – passing a couple of hundred feet below.'

Dick Rutan, Voyager world flight pilot.

BIBLIOGRAPHY

Brookes, Andew, *Crash: Military Aircraft Disasters, Accident and Incidents* (Ian Allan, 1991)

Brown, Captain Eric 'Winkle', *Wings on my Sleeve* (Weidenfeld & Nicolson, 2006)

Brown, Malcolm, *Lawrence of Arabia: The Life, The Legend* (Thames & Hudson, 2005)

Caidin, Martin, *Ghosts of the Air: True Stories of Aerial Hauntings* (Galde Press, 1994)

Cassidy, Brian, *Flying Empires Short 'C' class Empire flying boats* (Queen's Parade Press, 1996)

Gero, David, *Aviation Disasters. The World's Major Civil Airliner Crashes Since 1950* (4th edition, The History Press, 2006)

Gillespie, Ric, *Finding Amelia: The True Story of the Earhart Disappearance* (Naval Institute, 2006)

Grant, R.G., *Flight: 100 Years of Aviation* (Dorling Kindersley, 2002)

Gunston, Bill, *Fighters of the Fifties* (Patrick Stephens Limited, 1981)

Jarrett, Philip, *The Colour Encyclopedia of Incredible Aeroplanes* (Dorling Kindersley, 2007)

Lomax, Judy, *Hanna Reitsch: Flying for the Fatherland* (John Murray, 1988)

McKinstry, Leo, *Hurricane: Victor of the Battle of Britain* (John Murray, 2010)

Oliver, David, *Wings over Water: A Chronicle of the Flying Boats, Seaplanes and Amphibians of the Twentieth Century* (Apple, 1999)

Reese, Peter, *The Flying Cowboy: The Story of Samuel Cody – Britain's First Airman* (Tempus, 2006)

Regan, Geoffrey, *The Guinness Book of Air Force Blunders* (Guinness World Records Limited, 1996)

Richards, John, *A History of Airships* (The History Press, 2009)

Rivas, Brian and Bullen, Annie, *John Derry: The Story of Britain's First Supersonic Pilot* (Haynes, 2008)

Rosie, George, *Flight of the Titan – The Story of the R34* (Birlinn, 2010)

Smith, David J., *Britain's Military Airfields 1939–45* (Patrick Stephens Limited, 1989)

Taylor, Michael and Mondey, David, *The Guinness Book of Aircraft Records, Facts and Feats* (Guinness World Records Limited, 1992)

Telfer, Kevin, *Peter Pan's 1st XI: The Extraordinary Story of J.M. Barrie's Cricket Team* (Sceptre, 2011)

Wood, Derek, *Project Cancelled: The Disaster of Britain's Abandoned Aircraft Projects* (Tri-Service Press, 1990)